OVERNIGHT CODE

THE **LIFE** OF **RAYE MONTAGUE,** THE **WOMAN** WHO **REVOLUTIONIZED NAVAL ENGINEERING**

Paige Bowers & David R. Montague

Lawrence Hill Books

Chicago

Published by Lawrence Hill Books
An imprint of Chicago Review Press Incorporated
814 North Franklin Street
Chicago, Illinois 60610
ISBN 978-1-64160-259-4

Library of Congress Cataloging-in-Publication Data
Library of Congress Control Number: 2020947462

Interior design: Jonathan Hahn
All photos courtesy of David R. Montague

Printed in the United States of America
5 4 3 2 1

For anyone who needs to turn their obstacle into
a challenging situation

"It is not the mere presence of Black people that is the problem; rather it is Blackness with ambition, with drive, with purpose, with aspirations, and with demands for full and equal citizenship."

—Carol Anderson, PhD

"Each time a girl opens a book and reads a womanless history, she learns she is worth less."

—Myra Pollack Sadker

"I personally am willing to take it all the way. If I get shot then I just have to get shot. As long as you let whoever's pushing you up against the wall know that, you have a much better chance of winning."

—Jean Wheeler

"Oppressed people, whatever their level of formal education, have the ability to understand and interpret the world around them, to see the world for what it is, and move to transform it."

—Ella Baker

CONTENTS

FOREWORD

My mother, Raye, is at peace after a long life of many challenges and many triumphs. After her death in October 2018, countless friends, family members, and former colleagues gathered in Little Rock, Arkansas, to celebrate her. It was a send-off filled with countless stories, glorious singing, amazing tributes, and, of course, laughter. She was a woman who touched a lot of lives, and we've been fortunate to draw on the recollections of people who knew her in some way to introduce you, the reader, to a formidable woman who fought all odds to accomplish a goal she had from the moment she saw her first submarine at the age of seven.

Anyone who knew my mother knew that she had a way that she wanted things done. She was emphatic about staying on task and keeping others on task—especially when it came to any goal she wanted to accomplish, no matter the size. She always did so with a smile on her face. Yet, many can attest to how that gentle smile could be replaced with an archangel's stern visage when she dealt with people who did not want to do what was right. I think of her sitting in her hospital bed, telling me she was simply "tired and ready to go." But despite the fatigue that heralded her passing, she told me that I'd be speaking at her memorial and exactly what my topic would be. She felt confident that I would carry out her message with only that guidance.

First, she'd want me to say, "Respect the sacrifices made for you." The statement is not meant simply to honor the many who struggled to open doors for other people. It's something you have to think about in a larger context. Mom dealt with so many obstacles, yet always found a way to achieve despite the odds against her. For Mom, respecting the sacrifices is about what you actually *do* once you understand the implications of what someone did so you could attain your dreams. Her goal was always to make people's sacrifices for her more meaningful through her own actions and achievements.

My mother lived through the Jim Crow South. She lived through institutional and individual discriminatory practices that said—and, in many cases, still say—that a girl or woman should accept a reality of lower pay and lesser job opportunities. She lived through and overcame constant stereotypes about southern people. But she learned lessons from the people and places she encountered, always valuing education and training to constantly improve herself and accomplish her objectives.

As she had with so many others, my mother encouraged me to pay attention to her story and other people's stories to understand the importance of using my opportunities wisely. She advised me to pay attention to what was happening around me, strategize, set goals, and always be prepared with a Plan B when people tried to get in my way. Most of all, she told me not to let the ill will of others prevent me from paying things forward. That was a personal code of hers and it never failed to baffle her adversaries. In the end, Mom knew she wanted a clear conscience about doing the right thing, no matter what others said and did.

What my mother did for me, she also did for countless others, many of them complete strangers. That is why so many people considered her their other mother. After she returned to Arkansas, many people only knew her as my mother. They had no idea what she had accomplished and no clue about how many organizations she was active in, from the Links Incorporated and Alpha Kappa Alpha sorority, to the American Association of University Women and LifeQuest. She was a popular motivational speaker, a cherished mentor, and a fierce bridge player. It seems incredible to me that as accomplished and involved as she was, through it all she also managed to attend my school functions, organize and facilitate community events, socialize with her friends, and travel the world.

Raye at David's graduation from George Washington University in 1996.

She lived a full life, one of personal accomplishment but also one where she inspired others to live beyond the expectations they had for themselves or that others had for them. She saw herself as just one of many to do this work, and in her final years told everyone that despite her declining health, God kept her here for a reason: to inspire as many people as she could. This became more difficult for her as time wore on, but she once told me that she wanted to leave this world doing what made her happy. She found it a pleasure and an honor, and it gave her real purpose.

She was amazed that she had the opportunity to touch lives on such a grand scale. I can still hear her say, "Can you believe it? This is really happening." Everything in her life had come together, and I'm glad that she took advantage of the opportunity to spread her message globally in her final years and that so many people made that possible for her. Now, her fight is done, and she literally gave it her all.

Yes, she was an internationally recognized engineer who revolutionized the way the US Navy designed ships. But she was also a little girl from Little Rock, and my mother, too. This is her story. May it plant a seed in your mind and heart.

—DAVID R. MONTAGUE

I

JIM CROW

Raye Jean Jordan, age four, with her dog Bumpty.

1

Little Girl from Little Rock

Raye Means needed someone to give her a chance. On paper, the twenty-one-year-old college graduate should have been a shoo-in for any job opening that required a bachelor's degree and a flair for math and science. But it was 1956. Although Washington, DC, was on the verge of becoming the first majority Black city in the country, there were still racist attitudes there that made it hard for a young Black woman like Raye to embark on a meaningful career. She had already faced her share of obstacles back home in Arkansas, where she was treated as less than because of her gender, and then, because of her race, barred from pursuing the formal degree in engineering that she had wanted since she was seven years old. From that young age though, she began viewing obstacles as challenges that could be solved another way. Whenever Raye ran into trouble, her mother, Flossie Jordan McNeel, advised her, "Kick like the devil and holler for help."

So Raye came to the nation's capital prepared. She had set aside enough money to get her through a couple months of job hunting, but she was concerned about how long those funds would last. Her husband Weldon had been struggling to establish himself in the months since they had married, and she worried that things would become real tight real fast. As Raye blanketed the city with her resume, she made sure to hand a copy to her sister-in-law, Marge, who worked for the US Navy's David Taylor Model Basin, which was one of the largest facilities for ship design testing in the world. Marge brought Raye's resume to the personnel manager there, who quickly skimmed it. He noted her bachelor's of science degree—but missed that it was in business. He called Raye in for an interview, thinking she was a good candidate to work in the Applied Mathematics Lab (AML), which housed the large UNIVAC, a computer used to perform complex calculations for the government. The UNIVAC was a beast of a machine, a sixteen-thousand-pound behemoth with five thousand vacuum tubes that roared through one thousand equations per second.

"You are familiar with the UNIVAC, right?" he asked Raye.

"I am," she answered.

It was a lie. She had never seen a UNIVAC before; there were none in Arkansas.

Raye had also never typed a letter outside of the ones she pecked through in her college typing classes and at a part-time job for her alma mater's dean of instruction.* Yet she was so convincing during that interview that she was hired on the spot as a clerk-typist for the AML. The AML was staffed with engineers—all men—with degrees from Yale and Harvard. To Raye, they seemed to carry themselves as if being there was their birthright; they were the chosen ones, and she was not. Granted, this was not Raye's dream job, but it was a start, and it was a great comfort under the present circumstances. She had gotten in the door with a group of people who were doing what she wanted to do for a living. As far as she was concerned, there was no turning back now.

When you consider the world into which Raye Means was born in 1935, it is nothing short of miraculous that she was able to get out of

---

* Raye often said she had never typed a letter prior to her first job at the Model Basin, but that may have been a touch of hyperbole. Her college transcripts show that she was a well-trained typist, and her application for federal employment indicates that aside from her bluff about the UNIVAC, she was a solid entry-level hire.

Arkansas as a young adult and insert herself into the exact professional environment she desired. To hear Raye talk about her life in later years, it seemed as if she was blessed with an existence where everything simply fell into place for her, despite the roadblocks that she faced or that others attempted to throw in her way. The truth, of course, is always more complicated than that, before its rough edges are sanded down by time and triumph.

"My mother was the wind beneath my wings," Raye said. "I say this because if you think of where I was born and what color I am, I might not be able to tell you all that I'm about to tell you. And I have a lot of things to tell you, all because of my mother's love, faith, and support. Her advice has served me well in life; I've had to be scrappy to get what I want. After all, no one hands you anything. Especially if you're Black. Especially if you're female. You have to fight for it."

This hardscrabble work-your-way-out-and-up mentality was partly due to her mother's parenting, and partly due to the era. Although Raye was born at the lowest point of the Great Depression, many Arkansans had been struggling economically long before the stock market crashed in 1929. The Mississippi River Flood of 1927 inundated 6,600 square miles of the state with up to thirty feet of water, drowning homes, livelihoods, and valuable farmland. French poet Charles Baudelaire once said that one of the devil's great tricks is convincing you that he doesn't exist. But as the water lingered for five months, it was clear that the devil had no interest in hiding or playing games. He prowled about in broad daylight, looking for victims to devour. Residents flocked to Red Cross tents in search of shelter, and the devil offered them malaria, smallpox, dysentery, and typhoid instead. No federal aid came. When the waters receded, people were forced to return to what remained of their homes and start over with nothing.

Secretary of Commerce Herbert Hoover called the flood "America's greatest peacetime disaster." Perhaps this was the devil's intent all along—to destroy, and then to humiliate. Large farmers went bankrupt. Hoover attempted to get rid of the old plantation system by establishing smaller farms that could be owned and operated by resident farmers. His proposal floundered, and many Black sharecroppers fled north to look for other opportunities. Those who remained were still rebuilding when a drought turned the state to dust some three years later. Times were hard, crops no longer paid for themselves, and banks closed their doors left and right. Unemployment was at 40 percent, and more than half the people without jobs were Black.

As the state's environmental disaster was compounded by the nation's economic catastrophe, a young Black couple named Rayford and Flossie Jordan were running a restaurant called Ray's Radio Lounge about forty-five miles south of Little Rock in a town called Pine Bluff. Radios were a popular source of free entertainment, offering anything from dramas to comedy shows and news programs. Whatever you loved to hear, you could find it with a careful turn of the dial. For the Jordans, that was live music, and it was always crackling across the airwaves in their restaurant.

For a time, their life was idyllic. They had a restaurant and a house, which they had completely furnished on credit. Unfortunately, the Jordans fell prey to the financial hell that had engulfed everyone around them. When they got the final bill for their household furnishings, they couldn't pay it, and they lost everything they had, including Ray's Radio Lounge. They moved in with Flossie's sister in Hot Springs until they could find employment.

Flossie had an education degree and sought a position in some of the nearby schools, while Ray looked for any odd job he could find. It wasn't long before they discovered there was no work to be had, so they left for Little Rock, hoping to find opportunities. It wasn't easy there, either, but Flossie found a job as a waitress while Ray worked at the local zoo and restocked shelves in some of the local department stores. They loved each other dearly, and seven years into their marriage, Flossie gave birth to their first child on January 21, 1935, a daughter named "Ray," which was then spelled without the *e*.

"It became a blessing," Raye would later say. "That masculine name would open countless doors for me before anyone could see what I looked like."*

As soon as she became old enough to understand, Raye would learn

---

* According to her birth certificate, Raye was born Oscar Ray Jordan, but family called her Ray Jean throughout her girlhood. By the time she was in high school, she began going by "Raye."

that her looks—especially her brown skin—would determine the way she was treated by the White people who lived and worked outside of the neighborhood where she grew up. Racial segregation was enforced by law in the South, and upheld through the 1896 court case *Plessy v. Ferguson*, which mandated "separate but equal" facilities for Black people. But Black people had also been creating their own residential and business communities since the end of the Civil War, and many of them became thriving cosmopolitan areas. In Little Rock, West Ninth Street was the heart of the local African American community, and Flossie and Ray moved there with their newborn when she was just a few months old. When the Jordan family arrived there in 1935, it was full of Black-owned-and-operated grocery stores, candy shops, barbershops, law offices, newspapers, and pharmacies. On Saturdays, the sidewalks teemed with people running errands, meeting friends, having fun. There were soul food restaurants and barbecue joints and guys on bicycles selling hot tamales from hot tin cans. Families paid a nickel to see movies at the Gem Theater. Mothers and daughters strolled to the salon to get their hair done together. Fathers bought their sons a hot dog at Bobby's Hot Dog Stand. Buddies whiled away the afternoon over a couple of games of pool at Red's. Children clamored for pint-sized wrought iron seats at Dr. Frank's Drugstore, where they could get an ice cream sundae and a hug from Dr. Frank himself. Performers like Ella Fitzgerald, Duke Ellington, and Count Basie flocked to the area to perform at the Dreamland Ballroom, because they knew it was a Black-friendly venue. Ninth Street had its own rules and rhythm, and its Black residents didn't have to watch what they said or did as closely as they had to in the White part of town.

There was always someone to see and always something to do. Raye's father took her to boxing matches, which is how she developed a lifelong passion for the sport. She also never forgot about the time her father took her to see the wrestler Ralph "Wild Red" Berry, a diminutive middle school dropout with a great vocabulary, who was one of the most notorious cheats in the sport. How did such a little guy succeed in a big man's game?

Although she made that change official on all her legal documents as an adult, we refer to her as Raye throughout the book to prevent confusion.

"My strategy is that of compelling them to proceed from a state of bewilderment and complete uncertainty to a disturbing sense of inferiority, putting them thus in awkward, perplexing, and vexatious situations on the horns of a dilemma," he said. "This is possible through my great depth of intellect, integrity, heroic boldness, leonine courage, scholarly mien, and alert perception."

Because Berry was a rule breaker, he became something of a lovable scoundrel at a time when people needed to believe that the little guy could win. Watching Wild Red made for a fun outing, and with memories like these, Raye said she had a good childhood. She grew up in a rented four-room bungalow, with a refrigerator and a telephone, both of which were rare for the time. The telephone had a party line, which meant that there were three or four people in the neighborhood who had the same phone number. When someone called, there was a different ring for each person who shared the number. Those who shared the line weren't always good about taking turns. To get the other caller off the phone, a user would pick up and ask for a turn, or they would pick up, hang up, pick up, hang up, until the person talking got the point. With plenty of family living nearby, it's unlikely that the Jordans had much need for a phone. Many relatives were a short stroll away.

"It was like having our own little village," Raye said. "There were always relatives around, whether they lived nearby or had come up to Little Rock to stay with us until they got on their feet."

Family took care of each other in those days. One uncle gave up his bus station job to a brother-in-law who had children to feed. Raye recalled family get-togethers, full of her aunt Pet's salmon cakes and turnip greens and plenty of singing and dancing.

"Most of the time, I was the only child in the bunch, and I got treated like a little adult, which I liked," Raye remembered.

Although Raye lived in a bustling Black mecca, White racists almost destroyed that. Shortly after the Mississippi River flood ravaged the state in 1927, a young White girl named Floella McDonald disappeared and police speculated that she had been "snatched up by a Negro." A Black janitor named Frank Dixon found the child's body a few weeks later in the belfry of the church where he was employed.

Dixon, along with his seventeen-year-old son Lonnie, became suspects in the murder, and a large White mob wanted justice. Police quietly transferred the Dixons to a jail in another town to prevent unrest, but thousands of people gathered in Little Rock anyway, eager for revenge. The furor died down for a few days, then exploded again when a Black man named John Carter was accused of assaulting a woman and her daughter just outside of the city. Another mob formed and went looking for Carter. When they found him, they lynched him, shot him almost two hundred times, then set his corpse ablaze and dragged it to the intersection of West Ninth Street and Broadway. At least five thousand people rioted in the neighborhood for the next few hours, until the Arkansas National Guard arrived on the scene. Though the lynching and hostilities were condemned, Black residents feared for their safety and many of them left town for good. Many who stayed armed themselves just in case. Eventually, the woman who accused John Carter of assaulting her and her daughter came forward and told the truth: he hadn't done a thing to either one of them.

That was the last recorded lynching in Arkansas—not that there weren't unrecorded ones, or other types of attacks on Black people. The fear from this particular event lingered for decades.

"My aunt Pet lived in Little Rock during that time, and she used to talk about what they did to John Carter because she watched it unfold outside my great-aunt's house," Raye said. "I asked her why nobody did anything to help him. She told me they couldn't do anything because they were all so afraid for their lives. 'We peeked out the window and there was nothing we could do,' she told me."

In later years, Raye would often pass by the spot where John Carter met his end, thinking about what Aunt Pet saw and the fear and helplessness she must have felt.

"I did not know that kind of racial violence as a little girl, because by then, the neighborhood was pretty insulated from whatever might have happened outside of it," Raye said. "Although we were fair-skinned and some of us could pass for White, we knew better than to interact with 'White' relatives if they passed us on the street. If they reached out to us first, that was OK, because they initiated it. If it happened the other way around, then it could cause problems."

The problems usually started when you left the confines of the Ninth Street corridor. Black people weren't as welcome in other parts of town, and White people let them know it.

"You could go to a Woolworth's, for example, where they had a soda fountain," Raye said. "But I knew I wasn't allowed to order from it. My mother would go into department stores, but salespeople would not let her try on hats because she had her hair pressed and curled. They felt she would soil the hats if she tried one on and didn't buy it. There were little things like this, silly things like salespeople being finicky about waiting on us in a shoe store. My mother would try on shoes, then tell the salesperson she'd like to try on a different pair, and the salesperson would tell her she couldn't afford it. So my mother asked for the manager, and asked for the shoe she wanted, and she bought it, even though, yes, we probably couldn't have afforded it. You came to expect things like this, and mother always said it would pass. One thing I knew was that these stores were always happy to take our money."

When social slights like these were paired with the day-to-day uncertainty that the Depression brought, it could be overwhelming for some people. All across the nation, people were struggling, but Arkansans seemed to have it the worst. To outsiders, all Arkansans seemed to be poorer, dirtier, hungrier, and more backward. Within the state, Black Arkansans had a harder time getting jobs with some of the New Deal projects that were hiring in the state, because Whites didn't think they deserved that kind of pay. Black people also didn't get the same level of federal assistance that White people got, a fact that made daily struggles all the more disheartening.

"I don't know whether those stresses were the reason that my father turned to drink, but he became an alcoholic when I was a little girl," Raye recalled. "He wasn't abusive, but he had gotten so that he didn't go out with us or anything. He didn't provide. My mother, as much as she loved him, couldn't live like that. I remember one day that one of Daddy's brothers came to get him and bring him back to Mississippi, where he had family."

Raye was four years old at the time. Her mother told her that her father was "drinking up all the money" and she was going to divorce him. At the time, many courts awarded boys to their mothers and girls to their fathers in a divorce. Raye, however, was awarded to her mother because of her young age.

"I don't know whether my mother thought Daddy would ever get his head on straight, but she knew she would never let him take me away from her," Raye said. "As for me, I don't recall any sense of sadness that he wasn't around anymore. I felt my mother was doing what was best because he wasn't providing for us."

Raye wouldn't see her father again until she was thirty-one years old. In the meantime, her mother moved on with her life, enrolling in cosmetology school and bringing Raye to events and other happenings in their community. Relatives continued to come and go, and Flossie took in extra boarders to make extra money. One of them was a young man named Donnie Lee Lindsay, a sharecropper's son who moved to Little Rock to find his way. He had dropped out of school, but he reenrolled and was making ends meet as a delivery boy for Johnson's BBQ.*

"Aunt Pet was always on him about getting their barbecue sauce recipe, but Donnie would tell her that Mr. Johnson never let anyone see him make it," Raye said. "But she stayed on him, and told him to keep writing down whatever ingredients he saw until he got them all."†

Raye grew up surrounded by women like this, who wouldn't take no for an answer and wouldn't settle for less. Flossie had four sisters, all of whom divorced and moved in with her and Raye until they got back on their feet.

"Divorce is prevalent in my family," Raye later said. "The attitude was that if something wasn't working for you, get out of it. You don't have to be ashamed. Just move on with your life. My aunt Gladys married seven times. The first time she did it, she was sixteen years old, and she stayed with him for twelve days. She used to joke that she

---

* Lindsay became a bishop within the Church of God in Christ (COGIC) and started his own barbecue chain, Lindsay's Barbecue.

† Throughout her life, Raye compiled a small but well-chosen set of barbecue sauce recipes that she clipped from newspapers. In her collection was a recipe for a ketchup- and molasses-based sauce, as well as a dry rub, both concocted by David Cox, a Little Rock–based world champion barbecue chef who "reserves the right to make minor modifications from time to time." Her favorite BBQ spot in Little Rock was Sims BBQ, which makes some of the most ridiculously good ribs on the planet.

was going to keep trying until she got it right. She was with her last husband for thirty-two years."

Raye was close to these aunts, most of whom didn't have children, but treated her as if she were theirs. Gladys, who was the most independent-minded in the bunch, was a beautiful, fiery woman who instilled in Raye the need to stand up for herself and be capable of handling things on her own.

"She was the sweetest person you'd ever meet, but she'd cut your throat if you ever did the wrong thing," Raye's son, David, said. "It wasn't until I was older that I realized she walked around with a .357 Magnum in her robe."

Raye's aunt Angeline, who lived in the country, was the only sister with children, and Raye said all of those cousins resented her because they believed she got the most attention and the best of everything. It wasn't necessarily true. Raye did not have playmates her own age, so she said she spent a lot of time alone. For fun, she recalled she used to walk to Hubble Funeral Home and ask them if they had any new bodies that day.

"They'd tell me 'Raye-Raye, we don't have them dressed yet. You go out and play with the fish,'" she recalled.

Outside the funeral home there was a stone fish pond that had fragments of glass embedded in it. When the sunlight hit the colorful glass shards, they sparkled like fireflies as Raye waited and played.

"Then, I'd go in and sit and look at the bodies," she said. "I often saw the bodies before the families did, and I'd sit there with them for a while and go home. I didn't think it was odd at the time, and nobody said it was peculiar. I just knew that everybody in the neighborhood knew me and would look out for me, especially my mother and aunts."

Because of her living arrangement, Raye grew up understanding that women could do anything and everything a man could do. It was all just work that needed to be done. Flossie was a great electrician, Raye recalled, and she taught her daughter how to rewire anything in the house. Raye enjoyed the supposedly male work, especially because she was never a fan of dolls. She preferred trucks and trains and things of that sort, and would take them apart and reassemble them.

"Maybe that was the engineering side of me just starting to come out, but it was what I enjoyed," she said. "I remember hearing all the other little girls talk about what they wanted to do when they grew older, and they would say they wanted to cook and clean. I was never interested in that, even though people would say the things I liked were for men or boys. I was raised to believe there was nothing I couldn't do, so I believed it was the truth, because that's what my family told me."

Besides, she was a precocious child. By the time she was four years old, Raye could read and knew her numbers. She wanted to go to school, but public schools wouldn't take her because she was too young.

"Even if they had taken me, I probably wouldn't have been challenged," she said, explaining that the segregated schools that were available to her did not offer Black students as much opportunity to excel and achieve. Schools were separate, but certainly not equal, she said.

"White children had nicer facilities, harder classes, and better resources than we did," she said. "White teachers got paid more than Black teachers, and schools invested more in White students than they did Black ones. Education in Black schools reflected the jobs that were available to Blacks at that time. You'd learn some basic literacy, but you'd never be prepared to be much more than a laborer, or a housekeeper. Because of the system that existed, very few Blacks became lawyers or doctors."

Sensing she had a gifted and very self-assured child who needed to be challenged, Flossie enrolled Raye in St. Bartholomew's Catholic Church because it would take her despite her age. Although Raye's family was Baptist, she would become Catholic after becoming fascinated with the faith in catechism class. Because St. Bartholomew's was eighteen blocks from the family's home, Flossie couldn't take Raye to school every morning. She would tie Raye's bus fare to her wrist, walk her to the bus stop, and make sure she got on the bus safely.

"The driver always knew where to take me," Raye said. "So my mother trusted that I would be OK."

Flossie's desire to have her child get every learning opportunity that was available to her was rooted in her own upbringing. Her father, Thomas Graves, was a farmer who believed in education, even though he had no college degree. Graves owned his own farm in Moscow,

Arkansas, and he employed sharecroppers in his cotton fields. He insisted that all nine of his children go to school.

"In the country, schools didn't go very far," Raye said. "By the time my mother and her siblings got to high school, they had to go off to a college that offered them the classes they needed."

Raye's mother attended college and lived with an uncle, Dr. William E. O'Bryant, who had gone to medical school and was working as a pharmacist. He earned his degree in 1902. Flossie earned her degree in education in 1927, making her the second generation in her family to do so. Both attended what was then called Branch Normal College in Pine Bluff. Eventually, the school would name the campus bell tower after Dr. O'Bryant. There was great pride in that honor, and a sense of possibility within the family about what could be achieved.

"My mother would drill into me, again and again, that all I needed was to be educated, because that's what her own father drilled into her," Raye said. "What I didn't know back then was that I'd still have to fight hard."

Arkansas in the late 1930s was no place for a smart Black girl who dared to be different. All the same, Raye was determined to find her own way, riding public transportation to school in the morning and

The W. E. O'Bryant Bell Tower at University of Arkansas Pine Bluff is named for Raye's great uncle, who was the first in her family to go to college. Raye attended UAPB when it was called Arkansas AM&N.

walking home in the afternoon with twin brothers who were tasked with making sure she got back to her mother and aunts safely. The end-of-day arrangement only worked well for so long.

"One day after school, the boys couldn't find me waiting for them in my normal spot, so they rushed home to tell my mother, who panicked," Raye said. "She asked a friend with a car to help her drive around town so she could look for me. All of a sudden, she saw me walking down the sidewalk with my head tilted back, and I was whistling away. When she asked me what happened, I told her I was tired of going home the same way, every single day. I wanted to try something new. My mother tried to explain to me that it was safer to come home with the boys, but I told her I needed to try new things and see where they led. That was me, always wanting to investigate."

Although Raye might have been a handful, her family never got upset with her for these shows of moxie. No matter what doors were closed to Black people at that time, no matter what challenges Raye and her family faced, they had a tremendous sense of duty to each other, which probably came from Flossie's father, Raye believed. On the fence and gates outside Mr. Graves's farm, there were markings made by passersby that indicated the family would welcome anyone—Black or White—and give them food and a place to stay.

"My grandfather would see young men walking down the road near his farm and ask them if they needed something to eat," Raye said. "He'd tell them 'Come on in and my old lady will feed you.' and she did. My grandmother always set the table with an extra plate, because as she said, 'You never know when Jesus might come,' meaning you never know when you might have to feed somebody else. Sometimes those men would stay with my grandparents and work the crops, and when they'd get to the point where they wanted to get married, my grandfather would give them "a start," which was two chickens, two chairs, a table, and those sorts of things, just to help that young couple get going in life."

Thomas Graves also owned a general store, and each year he'd put up a Christmas tree inside it, because his sharecroppers weren't allowed to have one in their little shotgun homes. People who worked for Graves came to the store to buy presents for other people and put them under the tree. That way, the community could still celebrate the holiday by giving each other gifts.

"My grandfather had smokehouses too, so there was a period of time when he was just slaughtering and butchering," Raye said. "I remember being a little city slicker and visiting him on the weekends. I'd go into the smokehouse and they'd have these big sides of bacon hanging there, curing. That's how they did it back then. When you needed bacon, you'd just go into the smokehouse and whack off a piece of it. I never did help with the slaughter, because my mother told me that if I had seen what happened, I might not have eaten a lot of the stuff that I liked."

Raye's grandparents had a big garden around their house full of vegetables. Two of Raye's uncles lived down the road from the house and would send their kids over to the farm for vegetables and meat.

"My grandmother was a very kind woman, and she'd go into the smokehouse to get them a wedge of something before sending it down the road," Raye said. "She'd also fill her apron with fresh vegetables and send them down the road too. Mother said this was because they had so much. They were not wealthy by any stretch of the imagination, but compared to others, they did well. My grandfather was a hard worker, and though he wasn't formally educated, he was smart. He was also energetic and knew how to diversify, a trait that has definitely been passed down through the generations."

Raye recalled learning that the sharecroppers had children her age, all of whom had to work in the fields, just like their parents. They went to school, but only up until a certain point, she said, adding that they would work the land, and then grow up and have children worked the land just like they did.

"My mother raised me to know I had other options," Raye said.

As Raye's grandparents got older, her grandfather began having strokes, so Flossie and her sisters tried to convince their parents to come up to Little Rock so they could be closer to the rest of the family. Flossie's mother didn't need much convincing. She was happy to move. But her father insisted on staying down on the farm because he believed his sharecroppers would take care of him. All he needed was his shotgun and his bed, he said. It didn't take long for him to figure out that he needed more than that. His sharecroppers didn't take care of him in the way he thought they would, and one day he called his family in Little Rock to let them know he was ready to move.

"He called the house and got Aunt Pet on the phone," Raye recalled. "He said, 'Daughter, this is your Pa.' Aunt Pet was a bit of a devil, so she decided to have some fun with him. She said, 'Who?' And he yelled, 'Your Pa!' She said "Who?" again, and he told her to come down and get him. She brought him up and he became a part of our little so-called village. I remember my aunt Pet standing out on her porch and shouting 'Lemonade, lemonade! Good old lemonade! Made in the shade and stirred with a spade.' I can still see my grandfather slapping on his hat and charging across the street to get a glass."

2

The Submarine

Certain events have a way of splitting one's life in two. There is a before, and there is an after, marked by the changes wrought by that pivotal moment where one realizes that nothing can—or should— be the same again. For young Raye Jordan, the catalyst happened a world away from Little Rock, when a slender periscope pierced the moonlit surface of the Pacific Ocean at 3:42 AM on December 7, 1941. Its lens was trained on the US Navy base that was nestled in a lagoon at Pearl Harbor, Hawaii. Although the United States had declared a neutral stance in World War II, it had begun to supply Germany's enemies with armaments and ships. The Americans had also been building up their military presence in the Pacific in hopes of stemming the tide of Japanese expansionism throughout the region.

Although it was early in the morning, Ensign R. C. McCloy and Quartermaster B. C. Uttrick stood on the deck of the minesweeper USS *Condor* on lookout duty, some thirteen miles southwest of Pearl Harbor's entrance. McCloy had his binoculars trained on this unexpected viewfinder, some fifty yards from the port side of their ship.

"What do you think?" McCloy asked, before handing his binoculars to Uttrick.

"That's a periscope, sir, and there aren't supposed to be any subs in the area," Uttrick replied, as the sub quickly turned 180 degrees.

Uttrick signaled a nearby destroyer that was also on patrol, and both ships searched the area with sonar for an hour. They found nothing. At almost 5:00 AM, the protective nets to Pearl Harbor opened to allow both boats to enter. Neither vessel reported that they had seen a submersible, and the nets were never closed. Fifty minutes later, 353 Japanese warplanes roared toward that station from aircraft carriers in the area, ready to rain hellfire down on the ships and aircraft stationed there. An hour before those enemy planes arrived, the first of five two-man Japanese submarines entered the mouth of the lagoon.

The mini-subs were a last-minute addition to what was originally planned as a blistering airstrike on the Pacific fleet. Just seventy-eight-feet long, the battery-powered Type A Ko-hyoteki-class sub could hold two torpedoes that had double the explosive power of those carried by Japanese bombers. They could also creep through waters too shallow for larger subs. The Japanese navy viewed them as a secret weapon and decided to give them a combat test run in the Pearl Harbor assault. Several hours before Japanese planes began assailing the enemy, larger submarines would deploy their smaller cousins seven miles from the entrance to Pearl Harbor. The subs would maneuver into the harbor, and once the aerial attack began, they would surface and fire torpedoes at the American ships.

The bombs and bullets from above—and more than two thousand deaths below—would be what Americans remembered about the day that President Franklin Delano Roosevelt said would live in infamy, and which forced the United States to enter World War II.

The little subs would become something of a footnote. US ships struck four of them. A fifth, called HA-19, had a defective gyroscope and got stuck on a coral reef as it tried to enter the harbor. An errant shot from the USS *Helm* sprung the vessel free and knocked the crew unconscious until midnight, when they awoke to realize that they had drifted to the other side of Oahu. The duo tried to scuttle the sub, but the fuse wouldn't ignite. One of the men drowned, while the second, Kazuo Sakamaki, washed ashore, and was taken as the United States' first prisoner of war. The Americans also hauled in HA-19 so they could search it for intelligence before dismantling it and studying its parts. When they were done, HA-19's components were shipped to the Mare Island Naval Shipyard near San Francisco, where they were reassembled.

As shipbuilders reconstructed the Japanese sub, the United States' entry into the war helped some struggling Arkansans put their lives back together. New ordnance plants within the state provided people with much-needed jobs, and housing and transportation were built near those factories. At the height of production, at least twenty-five thousand workers, most of them women, produced millions of pounds of explosive material, and millions of tons of detonators, fuses, and explosive primers. The wages were still too low for some workers, who were forced to seek better pay elsewhere; more than 10 percent of the state moved away in search of more lucrative jobs between 1940 and 1943. Many never returned. Half the state's teachers, who were paid $700 a year, left their jobs to earn more than three times as much in the defense industry. Schools across Arkansas struggled to provide children with the education they needed to thrive.

Meanwhile, the Federal Bureau of Investigation had begun arresting Japanese immigrants who were identified as community leaders. In the two days after the Pearl Harbor attack, the FBI jailed nearly thirteen hundred priests, Japanese-language teachers, newspaper publishers, and organization heads, all of them men with families. For the next month, it continued searching Japanese homes on the West Coast, confiscating shortwave radios, cameras, heirloom swords, and explosives used for clearing tree stumps. Two thousand more Japanese immigrants were arrested by the US government as paranoia spread about national security. Newspaper headlines stoked fears with talk of Japanese untrustworthiness, spy rings, and fifth columnists working in concert with their mother country.

"Perhaps the most difficult and delicate question that confronts our powers that be is the handling—the safe and proper treatment—of our American-born Japanese, our Japanese-American citizens by the accident of birth, but who are Japanese nevertheless," W. H. Anderson wrote in the *Los Angeles Times* on February 2, 1942. "A viper is nonetheless a viper wherever the egg is hatched."

Two weeks later, on February 19, 1942, President Roosevelt signed Executive Order 9066, which gave the Secretary of War power to set up "military areas with which to exclude certain people." FDR wrote that a successful war effort "requires every possible protection against espionage and against sabotage." One month later, the president created the War Relocation Authority (WRA), which identified ten sites across the nation where more than 110,000 Japanese Americans could be held. All of the centers—or, more accurately, concentration camps— were located on federal or public land that was far from strategic war facilities and large enough to hold at least five thousand people. Two of those centers—Rohwer and Jerome—were built in southeastern Arkansas on bleak, marshy lands in need of clearing, leveling, and drainage.

When local officials announced that the state would be home to two of these facilities, many Arkansans were indignant that the enemy would soon be living right in their backyards. The outcry was met with justifications: the WRA would be spending plenty of money in the state, the Japanese internees would be growing much-needed food, and those internees would remain within camp boundaries. Although those living closer to Rohwer and Jerome had the hardest time with the news about the camps, they accepted their presence as a consequence of war.

In Little Rock, the *Arkansas Gazette* appealed for tolerance, reminding readers that 75 percent of the Japanese who were being "relocated" (a euphemism) were American citizens too. One reader wrote back disapprovingly, saying Arkansas had become a dumping ground for the unwanted. Others railed against what they perceived as favorable conditions in camps that had not even opened yet. The teachers who did not leave state jobs to work in the defense industry went to teach for better pay at Rohwer and Jerome. When a local congresswoman went to look into the matter, she was told that the teachers would be working in adverse conditions, and only for twelve months. By the time both camps opened that autumn, detainees moved their scant belongings into numbered and tarpapered A-frame buildings that they shared with at least 250 other people. Anyone who felt his or her living circumstances were too stark was cautioned against escape by barbed wire fencing and strategically located guard towers.

The relocation took a psychological toll on these prisoners. Many of them had supported the United States in the war. When the government decided that they were untrustworthy, some of these Japanese Americans renounced their US citizenship and sided with the Japanese government. Those who remained US loyalists were shocked that they were deprived of their homes and liberty, and sought ways to prove their devotion. Many young male internees who were born in the US volunteered for a combat unit that was comprised of others who were also American-born. Women also stepped up, some of them volunteering for the Women's Army Corps and the Red Cross.

Overall, as political scientist Morton Grodzins wrote, "The sentiment against the Japanese was not far removed from (and was interchangeable with) sentiments against the Negroes and Jews." Although several civil rights groups supported the fair treatment of the Japanese, many of them accepted the government's argument that the measures they were taking with these camps were necessary, legal, and appropriate. The racist undertones of these roundups and this paranoia were left unaddressed.

Prisoners working outside the fenced area of Camp Rohwer sometimes were brought back to jail at gunpoint by local residents who feared they were Japanese paratroopers. At Camp Jerome, a farmer who had just finished deer hunting came across three detainees who were on a work detail in the woods. He thought they were trying to escape and shot at them, wounding two of the men. When asked why he shot at the men while a White supervisor was present, the farmer told authorities that he assumed the overseer was trying to help the prisoners escape.

"A Jap's a Jap," Western Defense Command head General John L. DeWitt told the House Naval Affairs Subcommittee. "There is no way to determine their loyalty."

As Arkansas grappled with how to manage the enormous influx and incarceration of this ethnic minority, families around the state worried about relatives being called up to fight overseas. Raye's uncle, Willie Maurice Graves, enlisted in the army, and she said she didn't see him again until after the war. She and her relatives gathered around the radio to hear FDR's fireside chats for updates on the war's progress, just to get a taste of what Willie was facing on the battlefield. Raye said no one tried to shield her from the harsh realities of the conflict, and she enjoyed these evenings listening to the president talk about the world beyond Little Rock. For a girl who loved to explore, it was a way to imagine the many people, places, and things she could encounter in her future. Her present was focused on the day-to-day.

After school, Raye came home to Aunt Pet's lemonade and relatives sharing family lore. For a curious little girl who was coming to terms with who she was and grappling with who she might become as the world was seemingly coming apart at the seams, these moments were a perfect way for her to fade into the woodwork and learn by listening to the adults speak. Their stories formed a vibrant patchwork quilt, like the ones that the enslaved once draped over their so-called master's fences; the proud patterns came together in a way that not only looked beautiful, but also quietly pointed the way to freedom and a future.

"From those days when I would just sit back and listen, I learned that I could trace my ancestry back to 1608, when my Scotch-Irish ancestors, the Graves brothers, arrived on the second supply ship to land at Jamestown, Virginia," Raye said. "This was before the Pilgrims arrived, and before they even had states. My grandfather's name was Oscar Graves and he fought for the Confederacy; he was probably White, but he classified as mulatto. He married an African American woman named Winnie from Tennessee who was also Blackfoot Indian. My grandfather, John O'Bryant, fought for the Union and married a woman named Angeline Griffith who was part Black and part Cherokee Indian. So I am a daughter of the Confederacy and the Union. People look at me strange when I tell them that, but that's the way it is. People also ask me if I can take sides, and I tell them no, I can't."

That was her mother's side of the family. On her father's side, Raye said there was also Native American blood. Her great-grandfather was a slaveholder who had both Black and White children, all of whom he made sure were educated. When that great-grandfather died, he freed his Black children and left them a block of his plantation, mandating that it was not to be divided or sold until grandchildren came along. He wanted to give something to future generations, but he also wanted to be sure those generations could stand on their own without it.

Self-sufficiency. Education. Planning for the future. These weren't just themes that were drilled into young Raye's head about her present, they were time-tested values her kin had embraced and perfected long before her birth. No matter the tale, Raye noticed a common theme of love and respect for others, no matter who they were or where they came from.

"I grew up feeling loved and cherished by my family and neighbors," Raye said. "My mother always wanted to keep me in the limelight because she felt I had so much going for me. So she made sure I was included in things, because I guess I was a little different. She didn't want me feeling left out."

Nighttime changed that though, as the glories of Ninth Street weren't always suitable for little girls Raye's age. Sometimes the adults just wanted to go out dancing.

"They'd want to go out dancing, and they'd ask me if I wanted to come along," Raye said. "I always said yes, but they told me I needed to show them what kind of dancing I could do if I wanted to come. I would jitterbug and do all these other kinds of dances, and by the time I was done, I was worn out, and they'd tell me it was too late for me to come out with them. Someone stayed behind with me while they went out to dance all night."

Though she wasn't allowed to go out dancing with the adults in the evening, Raye's grandmother made sure she learned how to play the piano during the day. Raye's grandfather gave her mother, Flossie, the money to take music lessons while she was away at college, but Flossie decided to spend the cash on anything and everything else. When Flossie came home to visit her parents, she was surprised to find a brand-new piano sitting in the family's living room. Mr. Graves encouraged Flossie to play it for him, but all she knew how to do was find middle C.

"My grandparents wanted someone to play that piano, and that someone turned out to be me," Raye said. "Twice a week, my grandmother gave me twenty-five cents to go down the street to a music teacher named Ms. Collins. She was an oddball-type person who didn't socialize that much. Her world revolved around music and her son, but you never really knew anything about her or her family."

Although Raye said she didn't have a natural ear for music, she continued to take lessons from Ms. Collins because her grandmother told her she had to do it. Raye memorized all the music she was taught and started playing in recitals.

"Everybody thought I was going to be a concert pianist," Raye said. "That was not my forte, but music and math are aligned to engineering and people don't realize that. For a child like me, who was always treating things like a big puzzle that needed to be solved, it was much more interesting for me to see how all these notes on the pages came together to create a whole song."

Music was one of the things that drew soldiers from nearby Camp Robinson to Raye's neighborhood in the early evening. They were looking for fun and a bit of good food after a long day of basic training. They also had money to burn.

"The soldiers had to be off the streets by midnight, so people would let them come into their homes to play cards and laugh and talk because they didn't want to go back to camp," Raye said. "My family would welcome them, of course."

Raye's aunt Gladys rented rooms in the back of her house to some of the Black soldiers who needed a place to stay because hotels in the city were segregated. In time, Gladys found she could also rent to men who needed a short-term place where they could canoodle with prostitutes. She had a nice parking pad built in the back of her house, and her back door had a buzzer with a special ring that let her know she had a renter waiting. After Gladys let the patron in, she directed him and his companion through a beaded divider to their room.*

But the main goal was to help weary soldiers, especially after a March 22, 1942, incident where White military policemen beat Private Albert Glover, a Black soldier they had arrested for public drunkenness. After seeing the scuffle, Sergeant Thomas Foster, another Black soldier, confronted the officers about their unduly rough treatment of Glover. The military police then attempted to arrest Foster, but he

---

* David Montague said that after Aunt Gladys died, he and his family moved into her old house for two years when they first relocated to Little Rock. "It was a cute little place," David said. "But after a while some of her old customers started coming back around looking to borrow a room for a little bit. I told them no way."

resisted and ran. Foster was chased by military and local police, backed into an alcove in front of a Black Presbyterian church, beaten, and then shot three times in the abdomen. Foster died, and Abner Hay, the officer who pulled the trigger, was acquitted of the crime. Black community leaders lamented Foster's death, saying that he was "the highest specimen of military manhood training to make the world safe for democracy, that now, he will never know."

Raye's family opened their doors to soldiers to prevent future incidents like this. In gratitude, some of these men took up a collection for her so she could buy war bonds. War bonds were sold to finance a conflict that was now being fought on multiple fronts for an unknown amount of time. Private companies and government agencies built support for the effort with advertising that also bolstered morale. "If you can't go across, come across. Buy war bonds!" one advertisement read. "Keep him flying! Deliver us from evil! You buy 'em, we'll fly 'em!"

No matter the slogan, no matter the image, these ads were designed to connect Americans to their fighting men and make them feel like they were part of the battle too. On October 27, 1943, the US Treasury Department unleashed a new secret weapon in their bid for higher bond sales: the captured two-man Japanese sub, the HA-19. It had been reassembled since Pearl Harbor and fitted with folding steps, catwalks, and portholes for the viewing public. The sub embarked on a national tour that began in San Francisco. Admission was one dollar in war bonds for children and five dollars in war bonds for adults. All over the country, people clamored to climb down into what newspaper reporters called an "oversized sardine can." It was a reminder of the fateful day in December that started it all, and a perfect way to coax patriotic Americans to donate to the cause. Best of all: it was on its way to Little Rock.

However, war bonds and patriotism could only unify the country so much. The truth was that Black Americans looked to the rhetoric about fighting dictators who deprived people of their rights and couldn't comprehend how their fellow citizens didn't see the parallels between what was happening in Germany with its treatment of the Jews and what had been happening in America since African men and women were deprived of all of their rights and worked to death in the fields. Perhaps it was willful blindness. But Black Americans still joined their White neighbors in decrying the Nazi government's systematic abuse and humiliation of the Jewish people, hoping that their loyalty would result in equal treatment.

Five months before Pearl Harbor, FDR signed Executive Order 8802, which prohibited racial discrimination in the defense industry. It was the first federal action—but not law—that prohibited employment discrimination in the United States. FDR said that "the democratic way of life within the nation can be defended successfully only with the support of all groups." Civil rights activists had planned to march on Washington to protest discrimination in the military, but Roosevelt's action led them to call off the march. He said fair employment in the military would be good for national unity and morale.

Although patriotic fervor swelled across the country, some of the ordnance plants in Arkansas turned away Black workers simply because of the color of their skin. And the four thousand Black people who enlisted in the military that year discovered that while they were fighting to uphold freedom in Europe, they were battling a culture that treated them worse than White people at home. Those who had not been passed over by draft boards were relegated to noncombat duties such as maintenance and transportation. As troop losses rose, the military was forced to reconsider its stance on Black troops. By the end of the war, at least 1.2 million Black men and women had joined the military, serving as pilots, infantrymen, officers, nurses, and more. The world's greatest democracy fought the world's biggest racist with a segregated army. It was ironic, but it was also a sign of the times. Nothing, not even war, would force White Southerners to stop treating Black people as their inferiors.

"Black people were fighting for the rights of White people, but when we rode public transportation, we'd have to sit in the back of the bus, give up our seats to White passengers, or stand," Raye said. "Because of my mother's heritage, she was fair-skinned with beautiful bright green eyes and red hair. She could pass for a White person. She was even listed as White on her drivers permit. I was brown, but I never saw myself as different from her."

One day, Raye said she got on a city bus with her mother and a White soldier gave up his seat. Flossie said "God bless you" to him and sat down. As the bus driver drove off, Flossie pulled Raye onto her lap. When the driver looked up, he saw a young Black girl sitting on what he thought was a White woman's lap and stopped the bus.

He told Raye, "You give this White woman her seat."

Raye didn't understand what she had done wrong. This was her mother, and it made perfect sense to sit on her lap. Raye began to cry and Flossie started hushing her, before the soldier stood up and said, "If she can't sit here, then nobody else can."

Raye continued crying as she and her mother stood for the rest of the ride.

"When we got off the bus, my mother tried to explain the laws of the land to me, that Blacks were treated as inferior to Whites and forced to live, eat, learn, and ride separately from them," Raye said. "That's what this bus ride was about. She told me I had done nothing wrong, but if I wanted things to be different, I would have to vote someday to change these laws."

Not that voting would come easy. Flossie explained that Whites were also trying to restrict Black people at the ballot box.

"I remember her taking me with her to vote one day and telling me about the poll tax that everyone had to pay so they could register to vote," Raye said. "Those who couldn't pay the tax couldn't vote, and the rule impacted Blacks more than Whites. It also impacted Native Americans, women, and poor people, too."

Of course, there were loopholes to the rule. There were clauses that allowed any adult male whose grandfather or father voted in a year prior to the abolition of slavery to vote without paying the tax. There were also literacy tests and intimidation, any act that White people could use to keep Black voters disenfranchised.

"Although the Fourteenth and Fifteenth Amendments should have protected our rights as American citizens, it wasn't always that way," Raye said. "People tried to keep us in our so-called place in subtle and not-so-subtle ways. Although I may not have understood that as much when I was a little girl, the older I got, the clearer it would become to me."

On a blustery fall day in November 1943, Raye began to see a way out. She said her grandfather grabbed her hand and took her for a slow, ten-block stroll to the harbor in downtown Little Rock. The area was usually full of fishing boats bobbing in the Arkansas River. On this particular day, the HA-19 submarine that had been winding its way across the country stopped right in Raye's hometown, and crowds clamored to see it in person. Raye said she pulled up the collar on her jacket as they neared the crowd of people that surrounded the vessel. As she stood in line waiting to tour the U-boat, she said she thought it looked like a baby whale.

"My grandfather wasn't able to go inside, but he gave me permission to take the tour," she said. "When it was my turn, I climbed up the little ladder and down the hatch into the belly of this miniature submarine."

She recalled descending into darkness and being overcome by the distinct smell of sweat and diesel fuel. As soon as her shoes clanged against the catwalk, she was enchanted by what she saw. A tall man in a neatly pressed uniform led her through the vessel and told her all about it, saying it was launched by a larger ship that was lurking in the waters around Pearl Harbor on December 7, 1941.

"That was exciting to me, but probably of great concern to the adults who understood the realities of World War II," she said. "Still, I looked in awe at all the dials and switches. There was a thick silver steering wheel, copper tubes, brass fittings, and steel bolts. There was a panel with interconnected mechanisms and moving parts that looked like the inside of a watch that had been taken apart."

The guide pointed to a long metal tube that he called a periscope and invited Raye to look through it. When Raye peered through the viewfinder, she saw tree leaves, buildings, and cars far from the harbor. She also caught a figurative glimpse of her future.

"Wow!" she said. "What do you have to know to do this?"

"You have to know how to be an engineer," the man in the nice white suit said. "But you don't ever have to worry about that, little girl."

Though she didn't realize it at the time, the White man's comment was an insult, designed to make a young Black girl believe that she had no future in engineering. Still, Raye daydreamed about the knobs and devices on the slow walk back to her house, where she informed her mother about her future career.

"You have three strikes against you, Raye," Flossie said. "You're female. You're Negro. And, you'll have a southern, segregated-school education."

She looked at her daughter's hopeful face. "But," she added, "you can do or be anything you want to be, provided you have a good education."

The question was what that good education might look like, if her daughter wanted to pursue what would have been considered an unconventional career path for a woman or person of color. Flossie walked Raye to the library to find out more.

"I needed to learn math and science, and I had to be good at thinking outside of the box," Raye recalled. "My mother told me I could be whatever I wanted to be as long as I was educated, so I was going to educate myself in these subjects and become an engineer who worked on ships."

As Raye would learn, this path would not be easy or straightforward, but her mind was made up, and she took the first steps toward her goal.

3

Life in Pine Bluff

After the war ended in 1945, Raye's mother, Flossie, remarried. It is unclear when Flossie met William Henry McNeel, or how long they courted before they fell in love and became husband and wife. What is certain is that, while Flossie's first husband couldn't provide for the family, her second one definitely could. McNeel was an older gentleman, and one of the only two Black postal clerks in the state. With his job came a respectable, steady paycheck. Because of his work, McNeel moved Flossie and ten-year-old Raye down to Pine Bluff.

Situated an hour southeast of Little Rock, Pine Bluff is a delta town known and named for its high bluff of pine trees situated on the Arkansas River's south bank. The hamlet grew into prominence because of its cotton and timber industries, as well as its river commerce. However, the 1927 Mississippi River flood, the following drought, and the Great Depression hit the area hard. Cotton fields, farms, and small businesses were destroyed, forcing more than half the population in the town to seek some form of aid, whether from the government or a local church. Then, World War II brought a munitions plant to the area, providing more than ten thousand jobs that remained after the fighting ended. The area was booming, and its population was growing. For a new and cobbled-together family like the McNeels, it seemed like a place of great promise.

Although the relocation took Flossie from the support system of family and friends that had lived mere blocks away from her in Little Rock, it's easy to see why she might not have struggled too much with the idea of a fresh start somewhere else. After all, Pine Bluff was not unfamiliar to her. She had gotten her college degree there at what was now called Arkansas Agricultural, Mechanical, and Normal College (AM&N). She had also run a restaurant in the town with Raye's father, until the economy ruined them both.

Now, things were different. Pine Bluff would provide a change of scene and pace. For six years, Flossie had been putting herself through cosmetology school and working multiple jobs to support herself and Raye. Even with helpful family members nearby, it could not have been easy on Flossie to stay strong and keep pushing for herself and her daughter. It had to be mentally and physically exhausting. Not that Flossie would have openly complained. She knew she had a gifted child, and she did everything she could to nurture that. But Flossie also might have recognized that she needed to nurture herself, too. After all, she was a beautiful woman and far from past her prime.

McNeel gave Flossie the best life he could. He was able to buy them a nice house—white with a blue roof—on a hill in an all-White neighborhood.

"People would come to the door and they'd ask, 'Is the lady of the house in?'" Raye said. "And there I was, a little girl, and I'd say, 'I'm the lady of the house.' I loved to do that because I knew they weren't expecting it."

What Raye meant was that people weren't expecting a Black family in a White neighborhood. Although there were family members on Flossie's side of the family who were fair-skinned enough to pass for White, in Pine Bluff there could be trouble if they divulged their heritage.

"I had a cousin who was so fair-skinned he passed for Italian, and because he couldn't get a job in Pine Bluff, he went to look for work in Chicago," Raye said. "When he came back home, he rode the White coach on the train and got picked up by White cab drivers, but he couldn't have them drive him to his daddy's house in the Black neighborhood. So he'd have them bring him to our house. I remember answering the door one night and he told me, 'Get back, Raye Jean. Get back. Don't let him see you.'"

Raye said that one day her mother wanted to use the phone, and a teenaged girl who lived down the street was on the line. "Mother kept picking up and hanging up, and eventually the person who was talking to my neighbor said, 'Who is that picking up the phone?'" Raye said. "And the girl replied, 'It's that little shit-colored girl.'"

Although Raye and her family were not truly welcomed in their new hometown, they did make inroads in their own way. Raye's stepfather built Flossie a small beauty salon in their backyard, and she opened her doors to a largely White clientele.

"She was the only Black owner of a beauty salon that did White work," Raye said. "I can still see that neon sign out front—Flossie's Beauty Salon—and all the women coming in to get their perms and stuff. Mother had all the permanent wave machines and everything."

Raye worked as her shampoo girl, and she listened intently as her mother's clients gossiped about other people in town and current events.

"She knew about a lot of things before many of her friends did," Raye's son, David, said. "When you're talking to a hairdresser, you tend to open up about how you feel about certain things, and that gave my mother an understanding of what people in Pine Bluff were really like. There's some real vulnerability when you're around all those scissors and chemicals. So Mom saw that people weren't always what they led you to believe they were in public."

Flossie may have found it easy to settle in with a new husband, home, and business, but Raye struggled to find her place. She was unhappy about moving from Little Rock to a smaller, industrial town in the Bible Belt, and she faced starting fifth grade at a brand-new school without any friends at all. The family was living well, but Raye felt that nobody really wanted her around. She felt like all the things she had been told about women and girls achieving anything they put their mind to didn't apply here.

"She saw women around her in traditional roles and without the education or ambition to change those circumstances," David said. "She had been motivated by my grandmother from the earliest years in her life, and now she was watching girls and women not being encouraged to participate in some of the same things she was told she could do. Everywhere outside of the environment she had at home, she saw a response that was discouraging instead of encouraging. She didn't enjoy it."

She also did not like her stepfather.

"He didn't like me, either," Raye said.

Raye said her stepfather called her by her middle name—Jean—instead of Raye because her given name was a reminder of Rayford Jordan, Flossie's ex-husband. Raye refused to answer to Jean, and her stepfather was eventually forced to call her by her first name. McNeel also tried to force the family to switch to his AME Methodist Church. While Flossie switched from Baptist to Methodist for her new husband, Raye steadfastly clung to her Catholic faith.

"So every Sunday, I would go to mass in the morning, and then come back to go to church with them just so they would let me go to the movies in the afternoon," Raye said. "When they passed sacrament, I would cross my arms, as an indicator that I would not take sacrament in my stepfather's church. They would have to reach over me to get it. My mother realized I was unhappy with him, and that the only reason why I was in that church was because of her. So she told me to hang in there and I did."

Since 1910, Black families in Arkansas had been hoping that the state's mandatory education laws would eventually result in an equitable learning environment for their children. If school was the way to achieve success and self-improvement, then the powers that be were stacking the deck in white students' favor. Some Black men and women told themselves that segregationist policies were old and ingrained and too hard to change overnight. Gains were being made, little by little, so they told themselves it was important to be patient. Others were ready to fight the slow march of progress. An Urban League of Greater Little Rock study from the early 1940s showed that while Black children were one-quarter of all students attending the city's schools, they received one-seventh of school spending. The city spent more than $3 million a year on facilities for White students and barely more than $400,00 a year on buildings for Black children. Salaries for principals and teachers were in line with those trends. Black principals earned $1,340, while White principals brought home $2,099 a year. White teachers made almost as much as White principals, with a $1,216 annual salary. Black teachers eked out a living with $724 in pay. This was not unusual for the South at this time, but it was also not encouraging. In *Sue Morris v. Little Rock School Board*, Morris fought in the Eighth Circuit Court of Appeals to have a salary equal to White teachers. The court ruled that salary discrimination was unconstitutional. Despite that win, Morris lost her job.

The White South was only interested in maintaining its socioeconomic upper hand, especially in higher education. A 1938 Supreme Court ruling, *Gaines v. Canada*, said that states that maintained segregated schools needed to provide comparable training for Black residents. Despite the ruling, progress toward that end was slow. After the war, however, there were Black college students who were ready and willing to push the issue.

One of those students was Silas H. Hunt, a World War II veteran who put himself through Arkansas AM&N working multiple jobs so he could earn a bachelor's degree in English. An excellent student, Hunt was set on attending law school at a university that would accept him, regardless of his race. Originally, that school was going to be University of Indiana, but when he heard about an AM&N classmate who was battling to attend University of Oklahoma's College of Law, Hunt decided that he, too, would test the rigid segregation that prevented talented young men and women from earning a degree closer to home. He applied to the University of Arkansas School of Law.

Southern schools that had denied Black students admission had weathered their share of negative press. Recognizing this, the University of Arkansas said it would admit qualified Black graduate students, but not undergraduates, making it the first White Southern university since Reconstruction to do so. U of A admitted Hunt to its law program on February 2, 1948, and he began taking segregated classes in the basement. Although Hunt had begun to pave the way for other students like him, tuberculosis would claim his life and prevent him from getting the degree that he fought for the right to earn. Still, others began to follow in Hunt's footsteps.

At the national level, President Harry S. Truman extended the wartime Fair Employment Practices Committee that monitored discriminatory hiring practices of Blacks in government and defense industry jobs. Discrimination was a big point of contention for Truman, as he saw after the war that foreign powers did not want to discuss human rights with a nation that treated Black citizens as inferior. Americans wanted to get back to their normal lives, but Truman wanted to address the "moral dry rot" at the heart of the country. When Truman recommended civil rights legislation, he found that only 6 percent of Americans would accept it.

Truman became the first president to address the National Association for the Advancement of Colored People (NAACP). Speaking to a crowd of ten thousand in 1947, Truman said, "The only limit to an American's achievement should be his ability, his industry, and his character." A few months later, his civil rights commission produced the report "To Secure These Rights" which argued again for civil rights legislation. Truman called for a federal law against lynching, stronger protections of the right to vote, laws against poll taxes, the establishment of a Fair Employment Practices Commission, and an end to discrimination in travel by bus, train, and plane. He also called on Congress to act on claims by Japanese Americans of being taken from their homes and forced into internment camps during World War II. He took a stand, but the country was not quite ready to take these progressive steps with him. The Democratic Party was divided between those who supported segregation and those who didn't, but Truman kept on, determined to make progress. After his upset reelection victory over New York governor Thomas Dewey in 1948, Truman brought forth his "Fair Deal," which included civil rights legislation for African Americans.

"I shall fight to end evils like this," Truman said of racism. He knew his battle would be better for the country in the long run, but after the struggles of wartime, White Americans weren't ready or willing to take on any skirmishes of this ilk. On the surface, the country was peaceful, prosperous, and full of promise. White people wanted to settle in and enjoy themselves and ignore the inequalities around them.

———⁕———

As the only Black child in a neighborhood full of White kids who went to an all-White school, Raye struggled. It did not help that her White neighbors were envious of her family because her stepfather made better money than they did. All of this reinforced the sense of loneliness Raye felt when she moved from a city full of people she knew to a town where she knew no one.

Although Pine Bluff High School was situated behind Raye's house, it was for White students only. Raye had to walk alone to Merrill High School, a K-through-12 institution that had served African American students since 1886. Named for philanthropist Joseph Merrill, a White New Hampshire man who donated money to Black families in Pine Bluff so they could convert a two-story red brick home into a school, Merrill had become known as a top-notch public educational facility with progressive teachers who made the most of their low budgets.

"My teachers in Little Rock were all White nuns," Raye said. "I had never had a Black teacher until I started at Merrill, and never had a White teacher again until after I graduated from college. But I tell you, I had some dynamite schoolteachers. The teachers at Merrill were just so bright."

Raye said all of her teachers had college degrees, but when they wanted to pursue advanced degrees, they weren't allowed to do it in the University of Arkansas system, despite the fact that the school allowed qualified Black students to pursue graduate work there. Instead, the state gave these teachers a stipend to get an advanced degree elsewhere, and many obtained degrees at places such as Columbia University in New York. When these teachers returned to Pine Bluff to work, they shared what they had learned with Merrill students. This was significant because Merrill got the old, outdated books from the White schools when those schools moved on to newer editions. The texts would arrive beat up and marked up, Raye said, but the teachers would make the best of it. Eventually, however, these teachers asked parents to invest whatever they could in getting the best and latest textbooks for their children.

"Our parents dug into whatever they had so our teachers could order what they wanted for us," Raye said. "So they taught us with the newest books, the newest technology, and it put us head and shoulders above the kids at the White school. The White kids in my neighborhood would come to me because I knew so much."

In later years, Raye would joke that these teachers set them up with master's degrees at an early age. "Our teachers built us up, told us we would go to college and achieve great things," Raye said.

Perhaps some of these teachers knew of the groundbreaking studies being done by Kenneth B. Clark, a New York–based psychologist and social activist who devoted his life to studying the impact of racism on children. In the late 1930s, he and his wife, Mamie Phipps Clark, conducted an experiment that poignantly illustrated how segregation and racism were harming Black children's psyches. Using two dolls—one Black, the other White—the Clarks asked African American children which of the dolls was good or bad, which one they preferred, and which one they identified with. Study after study, in the North or in the South, most of the children chose the White doll.

"The results of our studies were indicative of the dehumanizing, cruel impact of racism in our allegedly democratic society," Clark wrote. "These children were . . . seeing themselves in terms of the society's definition of their inferior status."

Clark found disturbing evidence of this when he visited rural Arkansas and tried the doll test with an African American boy. When he asked the boy which doll was most like him, the child pointed to the brown one and matter-of-factly used the *n*-word to describe the doll—and himself. Clark found the child's response as upsetting, if not more so, than that of the children in Massachusetts who refused to answer the question or cried and ran out of the room. Raye's teachers understood that they needed to go beyond the average expectation of teachers to reverse this disturbing trend. That's not to say that their work could rid Pine Bluff of racism, but it definitely went a long way toward showing a generation of Black children that they could overcome it by dreaming big, working hard, and never giving up.

Raye said that it was hard to remember that advice from time to time. On her first day at Merrill, she remembered walking into a classroom where it seemed like nearly all the desks were taken. She walked to one girl and asked her if the seat near her was taken. The girl pointed across the room and said, "There is an empty desk over there."

Every day Raye walked past the White high school and elementary school to get to Merrill. She was a chubby youngster who only fit in adult clothing, she said, and because of that she didn't have as many outfits as other girls did because of the cost it took to dress her. She was self-conscious as a result, and said she often felt bad when she heard other girls whispering about her wearing "that same green dress again." She was also younger than most of her classmates. And the only time she saw these other Black children was when she was at school.

"My family didn't have a car, so when I walked home, I never saw another Black child until I went to school the following day," Raye said. "I was an outsider. No one came to see me unless my mother had a party for me."

However, Raye eventually befriended some of the White children in her neighborhood. "We would head down into Pine Bluff and go to eat lunch together, but White people and Black people could only sit on opposite sides of the lunch counter," Raye said. "So we'd do that, and we'd talk to each other across the counter and the waiters would walk between us. We were integrating things then whether they knew it or not."

But not all of the White children were welcoming and open. Raye recalled passing by some White kids in downtown Pine Bluff who told her to go back to her own country, which she took to mean Africa.

"I told them I was Native American, Irish, and Scottish, so which country would they like me to go back to?" Raye said. "And I told them to go back to their country, because they were in mine. They were shocked that I would stand up for myself. Most people would have turned tail and walked away. But I stood there and I dared them to say anything else about it. I knew my history and they didn't know how to respond to that."

Despite struggles like these, Raye never withdrew within herself. She poured herself into her studies and extracurricular activities; she joined the debate team, and she became a school hall monitor and business club member. She continued her piano studies and tried out for school plays. But there were no math or science clubs for her to join, a source of great frustration for the aspiring engineer.

Another source of frustration: none of Raye's peers seemed to understand what she wanted to do with her life. "I remember I was in eighth grade and people were talking about what they wanted to be when they grew up and I said I wanted to be an engineer," Raye said. "People laughed because they thought I wanted to drive a train. That's all they knew about engineering."

Raye's teacher, Irma Holiday, overheard the teasing and came over to comfort her. "She told me not to get upset about them teasing me," Raye remembered. "She told me to keep aiming for the stars, because at the worst case, I would land on the moon."

Holiday's advice stuck with Raye for the rest of her life. In the meantime, the school continued to reinforce that all students had potential; when celebrities came to speak at nearby Arkansas AM&N, they would also come to speak at Merrill.

"They brought all kinds of interesting people to our school," Raye said. "They brought the boxer Joe Louis, track and field star Jesse Owens, Mary McLeod Bethune, the von Trapp family singers. I remember [the von Trapp's] bus was parked outside our auditorium and we stayed late to see them perform. I mean, my school gave us concerts from the Metropolitan Opera. Sopranos would come and sing for us. Don't you think that was inspiring? No one realized that Black kids were getting dynamite stuff. They thought they were keeping us down, but these things, these special events, helped us have pride in ourselves. I mean, we knew we couldn't go to the White library in town, so they built a Black library right across the street from our school."

Raye continued to focus on her goals, and her family empowered her to stay focused on her schoolwork. Instead of getting a part-time job like most teenagers did, Raye was told she only had to concentrate on school.

"The family was concerned that if she got sucked into getting a regular check early, she might not continue her studies," her son, David, said. "They wanted to keep her on track."

By the time Raye reached ninth grade, she recognized that as good as her school was, there were gender-based restrictions on some of the classes she wanted to take. At Merrill, the rule was that girls took home economics, and boys took shop from ninth to twelfth grade.

"I wasn't interested in home ec., so I didn't really go to class," Raye said. "I never learned how to put a hem in anything. I remember one day the principal stood up in the auditorium when we were in chapel, and said, 'This little girl gets an F because she has not been attending home ec.' He called me out by name, and I stood up crying because I had never gotten a bad grade. So I called my mother and told her what had happened to me, and she came up to the school and talked to the principal."

Flossie told the principal that Raye had no interest in home economics classes. She asked the principal whether Raye could take shop, or other classes like it, if she passed the home economics exam.

"I had a photographic memory, so I didn't have to attend that class," Raye said. "I could read and remember and I could ace those tests. So I did well on the home ec. test, and I got to take calculus and so on instead. I still can't sew and cook. My mother never taught me to cook. She told me to get out of the kitchen and get good grades."

Raye's focus on what were seen as male classes and a male profession was not something her classmates could relate to, and she felt isolated and ostracized.

"It was hard on her emotionally," David said. "And some of her Black friends didn't understand why she didn't live where they lived. So she got it from both sides. She was always wondering where she fit in."

⸺ ✻ ⸺

As constrained as Raye may have felt, there were people in her life who were trying to show her the world was a big place full of opportunity.

At school, there was Holiday who told her to aim high and ignore people who laughed at her dreams. Teachers were always impeccably groomed—as she herself often was—showing students the importance of making a good first impression. They told students to make eye contact when they spoke to people and taught them how to listen carefully without taking notes. This was about more than reading and arithmetic. This was about life skills and about making one's way in the world.

Her family was also determined to give her an education beyond the classroom. Her aunts now lived across the country, and they would host Raye for periods at a time. Outside of expanding her horizons, these relatives may have felt the need to help Flossie after her second husband, McNeel, died when Raye was fifteen years old. One of her uncles was a dining car cook on the Missouri Pacific Railroad, and he used his rail passes to bring Raye on trips. She went to Chicago on her own one year for the July Fourth holiday, and to California sometime after that.

"I remember going to Chicago by myself for the first time and the train derailed," Raye said. "It was so hot that the track split. It never turned over, but it jumped the track. I didn't know what was happening, but my uncle's coworkers gathered around me and took care of me."

Although Raye was supposed to dine in the segregated car, her uncle would come get her before the White passengers came in and seat her right behind the galley kitchen. There were white linen tablecloths, silver utensils, and sparkling glasses that captured her imagination as she dined alone and felt like a queen. When she finished eating, her uncle would escort her out of the car before the White passengers arrived for their meals.

"I had great exposure to things," Raye said. "My aunts made sure of it. In Chicago, I got to go to the prestigious Club DeLisa, where Count Basie sometimes played. I remember they put me in high-heeled shoes and got my hair into an upsweep, so we could go hear live music. I saw a television for the first time on this trip, and I was so fascinated with it because we didn't have one at home."

Black families who had the means to afford train fares often took their families on jaunts like these to expose their children to other

Raye's Debonettes group in Pine Bluff, Arkansas. Raye is fourth from the left.

parts of the country. However, where Raye basked in the bright lights of Chicago and swam in the surf outside of Los Angeles, other families were not as fortunate as she was to afford and enjoy such extended stays. The journey itself was the highlight. Once they arrived at their destination, these families turned around and went right back home.

Raye was fortunate, and she knew it. She was grateful to have people eager to expand her horizons. In Pine Bluff, one such person was a local caterer named Hortense Jones—or, as Raye called her, "Aunt Hortense"—who took an interest in making sure that young Black women in her hometown grew up to have grace and poise. Aunt Hortense assembled a group of promising young women in town and called them the Debonettes.

"Aunt Hortense taught us social graces and other things that she had learned or observed," Raye said. "She taught us how to entertain, how to dress properly with a hat and gloves, which forks to use and which place settings to get. We had a formal dance when we were fifteen years old, and she told us to set the stage and the standard. She told us that if she taught us, we were to reach back and teach others, too."

Raye was coming into her own, and as she worked in her mother's salon, making small talk while she washed and rinsed local ladies' hair, she learned that she could be comfortable in any setting, with any person, no matter their color.

"Regardless of what you do, some people will assume that because of the color of your skin, that they're right and they're better," Raye said. "I learned that while you may not like me at first, I can and will break through those barriers."

After Raye's stepfather died in 1950, she and Flossie had a new barrier to consider: how the two of them would get by on their own, and, ultimately, how they'd afford a college education. Without McNeel's steady postal clerk paycheck, they began to struggle, and they had to rely on the kindness of their neighbors and friends.

"That was one of the reasons why my mother believed in saving later in her life," her son David said. "I don't think my grandmother had access to any fund that could have helped them out. But then again, I don't believe a lot of people did. There were no real long-term assets or financial advising about how to prepare for the long term. Having a place to live was the primary concern. After that, people figured they'd be OK if they had a roof over their head. Then, they could find a job."

Because Raye was studious and driven, she might have been a natural for a scholarship of some sort when it was time to apply to college. In the meantime, she was also a lively young woman with sparkling eyes and a megawatt smile. She said she had various beaus in high school, none of whom attended Merrill. However, she did not have a boyfriend when the Debonettes announced a formal back-to-school dance in the autumn of 1950. Flossie, always wanting Raye to be in the thick of things, asked a neighboring family, the Means, if their son, Weldon, would take Raye to the dance. Weldon had fought in World War II, was six years older than Raye, and, as she described him, was "a really cute little boy." He was happy to escort Raye to the dance, and after that, they struck up what Raye called "a friendship."

"He had no reason to be interested in me," Raye said. "He was in college at West Virginia State and he was very smart. He majored in psychology and went to school around the clock. I only saw him during Christmas break, and for two weeks in the summer. Our courtship was mostly by mail."

Letters or not, Raye was a junior in high school, and all around her girls were getting engaged, flaunting their rings, and thinking about what might come next in their lives. Although Raye was determined to become an engineer, she was also young and had the attention of a bright and handsome young man.

"I think she had gotten starry-eyed," her son David said.

A handsome but introverted young man comes back from war, takes a bright young woman to a dance, and sparks fly. Was this relationship in letters true love or a temporary distraction? Only time would tell.

4

Aiming for the Stars

As Raye finished her senior year of high school, her social studies teacher spoke with students about the 1952 presidential election, a contest in which they would be ineligible to vote because of their age. However, the numbers of Black voters in the state had been steadily rising since a Pine Bluff–based lawyer and activist, William Harold Flowers, founded the Committee on Negro Organizations in 1940. After the war, Flowers teamed up with Black political, civic, fraternal, and religious groups to increase the number of registered Black voters from 1.5 percent to 17.3 percent. The *Arkansas State Press* called Flowers the founder of a movement, but the truth is that he was the face and force behind an undertaking that had been gaining steam since the war, when Black soldiers fought abroad for a freedom they didn't have at home. No matter what tricks were pulled at the polls, Flowers showed that there would soon be no denying African Americans their say in civil rights measures. They had a voice that was growing in power, and were ready to use it. In the meantime, Raye's social studies teacher, Ella B. McPherson, armed her students with the knowledge they'd need to wield their future political clout well.

Since 1947 there had been a movement to draft the five-star general Dwight D. Eisenhower as a presidential candidate. The Democratic Party tried first, in part because of the Southern Dixiecrats, who became disillusioned with President Truman's support of civil rights. Eisenhower rebuffed those advances, saying he would not identify himself with a political party, accept a nomination for public office, or participate in a partisan political contest. It would be against military regulations, and the last thing the former Supreme Commander of the Allied Expeditionary Forces would want to do would be to undermine the strength and legitimacy of the armed forces by stating his preference for a political party.

The furor for his candidacy died down after that declaration, but it surged again prior to the 1952 election, when Americans became frustrated by the seemingly endless Korean War. The United States had been concerned about the spread of communism since the end of World War II; the Korean conflict was a prime manifestation of this fear. The paranoia was so dire that there were even worries, stoked by Senator Joseph McCarthy, that communist agents were infiltrating the government. Who would have the strength to bolster America's status as a world superpower in the face of this threat? Americans wanted peace, prosperity, democracy, lower taxes, and varying degrees of civil rights. Who would step forward to ensure the status quo? At a time when Raye and so many of her classmates were heading out into the world, these were heavy questions to ponder.

"One of my classmates told Ms. McPherson that he would not need to know this stuff because he was going to be a farmer and wouldn't need it," Raye said. This was unimaginable to Raye, who didn't realize until her class practiced for graduation that not everyone planned to attend college like she did, and not everyone had the same life goals she had. Some were headed into the armed forces, while others said they were going to work in Milwaukee, Chicago, Detroit, Cleveland, and various other locales.

"I had been so sheltered and brainwashed that I did not know we had an option," she said.

On the other hand, this is the same young woman who decided at age seven what she wanted to do for a living. Her mother supported her mission, and because of that, Raye didn't seem to question her choice. As a matter of fact, she stressed that it was right for her, all the while noting that more traditional female pursuits were not. Perhaps in the single-minded pursuit of her own goals, Raye lost sight of what her peers wanted to do, or even *had* to do with their lives. She wouldn't have been the first teenager willfully oblivious to other possibilities as she forged a new, post–high school life for herself.

Raye's current chapter was coming to a close. In her senior memories album there were the usual good luck wishes and inside jokes that one finds at the end of a school year. Teachers wrote that success was hers, and that she had "the ability to do many things. Do not miss this opportunity." Another admonished her to "give to the world the best you have and the best will come back to you." There were newspaper clippings, class photos, and a doodle of Raye and her friends driving off into the sunset with the phrase "Just married to the future" on the back of their car. Although her beau, Weldon, was not in the sketch, he did sign the book, writing, "Congradulations [*sic*] to the sweetest and most affectionate 'chic' [*sic*] I know."

Graduation day came, and one of Raye's classmates decided not to attend. Raye said that the young man later admitted to sitting on his back porch and crying when he realized that his peers would be marching in to "Pomp and Circumstance."

It was a poignant day, Raye said, made even more emotional by the shared realization that as the class of 1952 stood there in their caps and gowns, it might be the last time that many of them would see each other. Raye, that awkward fifth grader who felt unwanted when she started at Merrill High School, had blossomed into a young woman who felt like she was part of a greater family of likeminded souls. Before she and her classmates walked across the stage to receive their diplomas, the school's choir broke out into a rousing rendition of the spiritual "Ain't That Good News":

> I've a crown in the kingdom
> Ain't that good news
> I've a crown up in the kingdom
> Ain't that good news
>
> I'm gonna lay down this world
> Gonna shoulder up my cross

Goin' to take it home to Jesus
Ain't that good news

The class motto was "Victory is ours forever." At least it was the case on this particular day, May 23, 1952, as Raye and her classmates stood there, diplomas in hand, and bid each other tearful farewells. Though they couldn't know the nature of their fight, there would be more battles, more victories, and occasionally, more heartbreak ahead.

———∞∞∞———

Raye Jordan was coming of age in a world that had changed dramatically in the decade since the US government secretly launched the Manhattan Project, a venture to develop the first atomic bomb. Although physicists who worked on the mission were morally opposed to the use of the destructive beast they were creating, President Truman believed that if it were dropped on Japan, the destruction it wrought would persuade the Japanese to surrender without an invasion of American troops. On August 6, 1945, an American B-29 bomber dropped the first atomic bomb on Hiroshima, which instantly killed almost eighty thousand people. Three days later, another plane dropped an a-bomb on Nagasaki, ending the lives of some forty thousand more. On August 15, Japanese Emperor Hirohito announced his country's unconditional surrender, noting the overwhelming power of a "new and most cruel bomb." These explosions showed the world that nuclear arms seemed to be the way to best guarantee a country's safety in uncertain times. Although many nations began their own nuclear development programs, the United States and Soviet Union were farthest along in the race. However, this quest for global safety seemed only to guarantee more destruction. The more countries with nukes, the more likely it seemed those weapons would be used, with disastrous effects.

Then, US scientists began a quest to develop a "super-bomb" that could fight the Communist threat with even more catastrophic force. Where the atomic bomb's power resulted from the energy released after a heavy nucleus was split, this new bomb—the hydrogen bomb—

brought doom with the energy released when two light atoms combined. Developing this weapon required computations well beyond the capability of the average calculator. Mathematicians had been in the process of developing a more modern machine with vacuum tubes that empowered it to do high-speed computing. Some of them called it a human brain, while others called it what it truly was: a computer.

The Electronic Numerical Integrator and Computer, commonly called ENIAC, took up fifteen thousand square feet and used eighteen thousand radio tubes to perform thousands of calculations a second. Its creators, John Mauchly and Pres Eckert, had invented a machine that extended man's ability to do complicated math problems, a feat that would be useful in the fast-changing postwar arms industry. Funded by the US Army, the world's first all-electronic digital computer was one thousand times faster than other calculating machines.

ENIAC's first job was to calculate ballistic trajectories for the army, and a team of six female "computers" was hired to program the machine. Kay McNulty, Betty Jennings, Marlyn Meltzer, Fran Bilas, Ruth Lichterman, and Betty Snyder had all been calculating ballistics trajectories with a mechanical calculator before they were hired to work on ENIAC. Access to the computer required a security clearance, and until the women received one, they learned how to operate the machine's switches and cables via diagram. Ten months after they were hired, the female computers finally saw the ENIAC.

There were no programming languages yet, but the women—now called computer operators—were so adept at working the apparatus that they could fix it with greater ease than the engineers who had created it.

"You know how engineers like to create, but they don't like to have anything to do with the debugging," said Jean Bartik, who was then known as Betty Jean Jennings. "Well Betty [Snyder] and I got so we could figure out which vacuum tube out of eighteen thousand had burned out. We could run our debugging program and see if there were errors. . . . I tell you, those engineers loved it. They could leave the debugging to us and get to work on their next machine."

Without the operators, however, ENIAC wouldn't have been able to produce the calculations that would make the hydrogen bomb a reality. It was Snyder who learned to speak a computer's language, mainly because she was the one who created its earliest iterations. She was a former math student at University of Pennsylvania whose professor told her she'd be better off at home raising children. Although Snyder left the program to study journalism, which was seen as a more suitable female pursuit, mathematics still held a very special place in her heart. She earned her degree, and then returned to her first love, which brought her to ENIAC, and then to a crucial role in developing its successor, the Universal Automatic Computer, or UNIVAC, in 1952. UNIVAC gained notoriety for correctly predicting a landslide victory for Eisenhower in that year's presidential election. Traditional pollsters had called it for Adlai Stevenson, and the difference in human versus computer predictions was so dramatic that CBS News refused to read the UNIVAC forecast until Eisenhower had indeed won in a landslide.

Snyder married, became Betty Holberton, and was hired to run the programming research branch of the Applied Mathematics Laboratory at the David Taylor Model Basin in Carderock, Maryland. The naval ship testing facility was the only place large enough to house the US government's UNIVAC—and Holberton's imagination, which conjured other advances in a field that was becoming increasingly important. Through her work, Holberton would also pave the way for other bright, ambitious women like her—women who dared to flout convention and work in fields that were stereotypically and unapologetically male.

Holberton's credo was "Look like a girl, act like a lady, think like a man, and work like a dog."

Although Silas Hunt had begun the long, arduous process of integrating the state's universities when he was accepted to University of Arkansas School of Law, Raye Jordan would not be able to follow his path. U of A was the only school in the state that offered engineering degrees, but it did not admit minorities into any undergraduate program.

"Mother didn't have the money to send me out of state to attend college, so I went to Arkansas AM&N right there in Pine Bluff and got a bachelor of science degree in business, [but] I still took all these math and science classes I knew I'd need to become an engineer," Raye said.

Raye said she pursued a business degree instead of a math or science degree because that was what her school counselor advised. When asked why her counselor steered her in this direction, Raye said, "I don't know. That's just what they said to do." David Montague said that his mother told him that she followed the advice because if she couldn't work in the hard sciences, she wanted to be able to work in a field that could be just as exacting. Raye's mother and aunt Gladys were both entrepreneurs, so she believed she could become one too. As a backup plan, she obtained a secondary education license, "because everyone who graduated from college had one." It wasn't necessarily what she wanted to do, she admitted. But Raye knew there were opportunities for young Black women to become teachers, and she wanted to be sure she had options. Raye believed in having options, after all. On the bottom of a school transcript, she imagined what her future might look like if she wound up teaching. "I am best qualified to teach Business Education, Arts and Crafts, Psychology, and History Courses," she wrote, pointing to the classes in which she had some of her best grades. She made no mention of teaching math and science, perhaps because she harbored grander ideas and aspirations for herself in those fields, regardless of the degree she would ultimately earn.

As pragmatic as Raye tried to be about being steered away from her fields of interest, deep down she had to be frustrated. She grew up being told that she could do or be anything she wanted to be, and now a perfect stranger was steering her away from a future in science and math. Granted, college is often an adjustment for anyone, even stellar achievers like Raye. But in the beginning, so much felt beyond her control. Her stepfather had died, her mother was struggling to send her to school, and her boyfriend was living out of state. As much as she loved to learn new things, it's easy to see how Raye could have been distracted, overwhelmed, and even frustrated.

"She was just like everybody else, struggling to go to college and not knowing how she'd make it from one quarter to the next," said

Bonnie Dedrick, a college friend of Raye's. Although Dedrick was a freshman like Raye, she was a couple years older than most first-year students because she grew up picking cotton on her father's farm in Wilmot, Arkansas.

"We couldn't go to school until the cotton was out of the field," Dedrick said. "We'd be in the fields picking and we'd watch the white kids go by on the bus to school. When the cotton was picked, then we could go too. That's the way we did it until we finished at AM&N. We couldn't go straight through like Raye did. It was a struggle, but through it all, we made it. God was good to us."

Dedrick recalled seeing Raye walk toward town from her house on the hill each day. Sometimes Dedrick, who lived in an apartment with her sister and brother-in-law, met up with her on that stroll, and they would laugh and chat on the way to campus.

"I knew of Weldon, but I didn't know him personally," Dedrick said. "He didn't go to school with us, but Raye would fill me in about him."

Then, the duo parted ways and headed to class. Raye joined the debate team, but she soon had difficulty getting to campus at night because the buses didn't run late enough. Even during the daytime, she ran into trouble getting where she needed to go. Some days, when she was running late, she caught a bus to campus. Her mother gave her five dollars a week for transportation, but one day Raye had not gotten change for the ten-cent ride.

"I had five dollars and I remember the bus driver screamed and yelled at me and told me to get off the bus because he wasn't going to make change for me," Raye said. "I got on that bus every day and he knew me. Maybe he had a bad night. So I got off the bus and went into a drugstore to get change."

By the time Raye walked out of the drugstore, the bus was gone, and it wouldn't be back for another thirty minutes. When it returned, Raye paid ten cents for the ride, and then realized all the seats on the bus were filled. Raye said the bus driver made her get off the bus and go around to the back door to get on.

"You think that's not destroying you?" Raye later said. "When you overcome those things, you wonder what those people think today when they see us excel in spite of the system."

Despite challenges like these, she threw herself into her studies as well she could, studying conservation with the author Alex Haley's father, Simon, during her freshman year.

"He would talk to us about Alex, who was in the Merchant Marines at the time," Raye said. "And he told us how he was writing a book."

That book would become *Roots*, a story that begins with a young man named Kunta Kinte, who was taken from The Gambia at age seventeen and sold as a slave, and ends with the six generations that came after him.

"When we didn't want to have class, we'd get him to talk about his children," Raye said. "He'd start talking about how Alex went for an interview at *Playboy* and when they asked to see him type, he just had that typewriter singing. And then there was his son George, who became a state senator from Kansas. President Clinton would later appoint him ambassador to The Gambia. There was also Julius, who was a naval architect, and Lois, who was a student at West Virginia State. His stories about his children were a great inspiration to us, and showed us that we could do these things too."

<hr />

On Friday, February 13, 1953, Raye was walking with Bonnie Dedrick and some other friends to an 11:00 AM gym class. To get to that part of campus, Raye had to cross a busy street, which had a police officer there to stop traffic so students could cross safely.

"That day, I remember the traffic stopped except for this one little truck that came through," Raye said. "The driver had just bought it that day for fifty dollars, and it had no brakes, and it hit me in the right hip and broke my leg."

Raye's shoes and books were thrown everywhere. Onlookers tried to pick her up, but she was in a lot of pain. The police officer sent for a gurney, carefully laid Raye on it, and then brought her to the infirmary while someone went to tell her mother what had happened. Raye was transported to the hospital by ambulance, which sent her home after determining that she had a broken leg.

"There was a Black physician, a Dr. Lawler, who from school, and he was not allowed to practice at the local hospital," Raye said. "He did what he could to help. My mother heated sandbags to keep the swelling down, so Dr. Lawler could come back and set my leg. No X-rays were ever done to see if the cast was put on properly, but as a result I was disabled."

Due to the accident and Raye's injury, the University of Arkansas system paid for the rest of her AM&N tuition, which classmate Rosenwald Altheimer said was about forty to fifty dollars an academic quarter. Altheimer, a schoolteacher's son who was one of five boys, viewed tuition as "dirt cheap." Not everyone saw it the same way, however. Flossie and Raye were barely making the payments until the accident happened and changed everything.

"It was a blessing to Raye," Dedrick said. "We all had our struggles, but she had tried to get into [University of Arkansas] Fayetteville and they wouldn't let her in. Then, she went to AM&N and she wasn't sure whether she'd have the money to finish. But this accident happened and she was able to go to school for free. Even if it hadn't happened, there was nothing that was going to stop her."

Raye's son, David, once asked her what she would have done about school if she hadn't been hit by the truck and able to finish school tuition free. "She said she didn't know," David said. "There was no real discussion about how they'd pay for school, no talk of financial aid, or getting a loan, or grants, or anything. She said that maybe she'd have gotten a job and gone to school part time, if it had come to it, but there was no real plan, prior to that accident, and that's scary. There she was pushing and pushing and my grandmother had nothing but goodwill, trying to encourage my mom without knowing what was going to happen next."

Because Raye's leg was not set properly, it would always be weak. She didn't have a noticeable limp, but she would never run and always tired quickly, which was hard for someone who loved dancing as much as she did. She was young and active, and although her leg healed, it would never be the same.

After Raye recuperated and her stay at AM&N was secure, she pledged Alpha Kappa Alpha sorority, finding in it a network of women who brought out the best in each other. Founded in 1908 by nine women at Howard University who believed that college-educated Black women could represent "the highest . . . of everything that the great mass of Negroes never had," their credo was to be "supreme in service to all mankind."

"This was a privileged group and in order to be in it, you had to be quite accomplished," Raye said. "You had to have excellent grades and they used to say that you had to be attractive and things of that nature, but that wasn't necessarily true. However, if you looked around at the people who became sorority sisters, they were usually good-looking girls."

Because Raye didn't live in a dorm, she had to come to campus at night in order to participate in Hell Week, the ritualized hazing that marks a pledge's entry into Greek life. It was called "being on the line" and a fellow pledge was called a "line sister." One night, Raye said she and her line sisters were taken to a cemetery where they were told to get on their knees, bang on a crypt, and scream, "Please, Mr. Dead Man, may I come in?"

"We were so tired that there was a slab, and I crawled up on it and went to sleep at midnight in the cemetery and they're calling on us," she said. "They couldn't find me. They were crying out for me, 'Raye! Raye!' because they knew they couldn't go back to campus without me. They found me asleep on that slab and never took me back to the cemetery again."

The late-night exploits could make for a long day in class, she said. "I remember I was in economics class with a fellow pledge and both of us looked groggy from the night before," Raye said. "My professor told us, 'Y'all go back to the dorm and get you a nap and don't come back to class until you come off that line.'"

If pledging taught Raye anything, she said, it helped her recognize that she could "hang with anybody" from any walk of life, and endure any challenge, silly or serious. Her sorority affiliation and friendships were something she would treasure for the rest of her life. She proudly showed off various keepsakes and jewelry she owned that were related to the sisterhood, her favorite possession being a thick AKA cardigan her aunt in Chicago had made for her.

"It cost $24, which was a lot of money at the time," Raye said of the sweater, which had green stripes on the sleeves and the sorority crest over her heart. "But this thing has double-lined pockets and weighs about fifteen pounds. You weren't going to be cold in it. That was for sure."

But you would be cool—and part of something bigger.

As Raye fought her way through her first two years of college, a landmark school integration case was barreling its way toward the US Supreme Court. In 1951, a man named Oliver Brown filed a class-action lawsuit against the board of education in Topeka, Kansas, after his daughter, Linda, was not allowed to attend the town's all-white elementary schools. Brown argued in his lawsuit that the schools for Black students and white students in Topeka were not equal, and that this segregation violated the equal protection clause of the Fourteenth Amendment, which mandates that no state can "deny to any person within its jurisdiction the equal protection of the laws."

The US District Court in Kansas heard the case and agreed that school segregation had "a detrimental effect upon colored children" that made them feel inferior. However, it upheld the "separate but equal" clause that had been a mainstay of American law since 1896, when the US Supreme Court held in *Plessy v. Ferguson* that segregation was legal as long as segregated facilities were of the same quality. The *Brown v. Board of Education* case went to the US Supreme Court with four other similar cases like it in 1952 and the justices were divided on how to rule. One of the judges, Fred M. Vinson, was firmly in favor of maintaining the separate but equal standard, but he died in 1953, just as *Brown* was about to be heard. President Eisenhower replaced him with California governor Earl Warren, who convinced his fellow justices to unanimously vote against segregation in 1954. He wrote that segregated schools were inherently unequal, so the plaintiffs were being "deprived of the equal protection of the laws guaranteed by the 14th Amendment."

The court never said how it wanted American schools to integrate, but it asked for further arguments about it. In the following year, it issued a second opinion on *Brown* that directed district courts and school board to integrate "with all deliberate speed."

While some schools followed the ruling, many southern districts did not. Things came to a head most notably in Little Rock, where school officials didn't seem to be in a rush to follow directions. Melba Patillo Beals was twelve years old when segregation in schools was struck down. She remembered her teacher being called outside of the classroom that day, and then coming back with a frightened, nervous look on her face. Her teacher erased the blackboard, told her class about the ruling, and then advised them that school would be dismissed early that day and that they should all hurry straight home, and walk in groups.

"Although she said the *Brown* case was something we should be proud of, something to celebrate, her face didn't look at all happy," Beals wrote in her memoir, *Warriors Don't Cry: A Searing Memoir of the Battle to Integrate Little Rock's Central High School.*

As Patillo left school, she realized she had left her math textbook behind, and she turned back to retrieve it. Her teacher refused to allow her back in the school, stressing the need to get straight home. She told Patillo she would excuse her math homework the next day. Patillo left, and was walking through a persimmon field on the way to her house when a white man called out to her, offering her a ride home. He had candy, he said. When Patillo refused to get in the car, the man chased her, slapped her, punched her, and attempted to rape her. As she fought to keep him off, the white man told her, "I'll show you niggers the Supreme Court can't run my life." A classmate heard the scuffle, came running with her backpack, and hit Patillo's attacker with it so that they both could run free.

After she went home and cleaned herself up, Patillo wrote in her diary: "It's important for me to read the newspaper . . . every single day . . . to keep up with what the men on the Supreme Court are doing. That way I can stay home on the days the justices vote decisions that make white men want to rape me."

Patillo wasn't the only one who was afraid. Other Black residents feared the ruling would destroy the gains they had made over time. They felt White residents would resent the court order, drag their heels to implement it, and react violently if it were enforced. Perhaps Whites needed more time, they felt. Quietly, Black residents knew that some new laws wouldn't easily change traditional practices.

Raye said the only time Black people in Pine Bluff were allowed to have an entire movie theater to themselves was when *Carmen Jones* came out in October 1954. A modern twist on the opera *Carmen* by Georges Bizet, the film starred Dorothy Dandridge and Harry Belafonte. Perhaps the cinema's manager didn't think White locals would be willing to watch a movie with an all-Black cast. Whatever the case, Raye and her peers liked not having to enter at the side of the building to climb three flights of stairs to seats in the balcony, where Black people were typically forced to sit. They could pay and enter in the front, sit anywhere they chose, and not have to navigate stairs with their popcorn and drinks. Of course, it was a limited gain, kept in place only for the duration of the movie's run. Everywhere else, people were dragging their heels about integration, whether in theaters, restaurants, or schools.

In the meantime, national newspaper and television reporters turned their attention to the civil rights struggle that was beginning to unfold. Their attention to the issues gave strength and power to a people who longed to be not just heard, but also understood and ultimately helped.

On May 27, 1955, Raye married Weldon Means in Washington, DC, with the understanding that she would come back to Pine Bluff and finish college.

"When Weldon graduated, he took a job in Washington because his sister was living there," Raye said. "I didn't really want to get married. I wanted an engagement ring because all the other girls were getting them. That's when he put pressure on me to marry him."

Weldon's sister, Marge Coleman, had a government job that paid $2,400 a year, which was not bad for the time, but she said she still felt like she was getting on her feet. She referred to her place as Grand Central Station, because she provided friends and relatives with a place to stay after they arrived in DC looking for work. Her brother, Weldon, would become one of a long line of temporary roommates.

"My father told me, 'You know your brother, Weldon, since he got his degree in psychology isn't doing anything with his life other than scheduling when he was going to make his next sandwich or meal,'" Marge said. "I was living in a basement apartment and my father felt that because Weldon wasn't doing anything but laying around, it would be good for him to get a start someplace outside of Arkansas. So Weldon moved in with me, and we lived in that shared space for a while, with only a room divider for privacy."

All that time, Marge thought she didn't have a good leg up, but apparently she was doing a lot better than other people. Eventually, Marge was able to buy her first house, and Weldon moved in with her. Raye came up in the fall of 1955 to spend time with him.

Marge had children back in Arkansas and was trying to earn enough money to bring them up to Washington. When she was finally able to move her kids to her home, Raye helped her care for them. It was a time when plenty of young people were trying to find their footing, sharing space and helping out when and where they could. Weldon was struggling, but when he asked Raye to marry him, she said yes. Funds, however, were an issue, so Marge hosted the wedding at her house, with sodas and snacks at the reception. She wanted to help her brother, and judging from the wedding photos, he was beaming, grateful.

Raye said it took two months for her to realize that her relationship with Weldon was not what she was expecting married life to be like. It is unclear what brought her to this realization and unknown whether Weldon felt the same way as she did at that time. Over the years, Raye never spoke about their problems specifically, perhaps because she had such a close relationship with Weldon's sister, Marge. It couldn't have helped matters, though, that Raye was going back and forth between DC and Arkansas for school. There was a growing rift between her and Weldon, and during one visit, the chasm felt so deep and unbear-

8/28/56

Raye's college graduation photo, 1956.

able that Raye decided she'd had enough. "So I packed up my stuff and went back home to mother so I could finish college," she said.

When it was time to graduate, Raye said she did not invite Weldon to the ceremony, most likely because of the marital trouble they were having. "We were outside taking pictures and one of Weldon's cousins said, 'Raye, look who's coming across the football field,'" Raye said. "And there he was, my husband, with my mother."

No one knows whether Flossie told Weldon to come work things out with Raye, or whether Weldon begged Flossie to help him. Whatever the case, the couple seemed to reconcile at least for a moment. The next day Raye returned to Washington, DC with Weldon. Deep down, perhaps she believed it was the best choice available to her. Racial tensions in the South were fraught. Emmitt Till had recently been brutally murdered; Rosa Parks had been arrested and thrown in jail in Montgomery, Alabama. The bus system boycott began soon after and brought a sense that something was beginning to shift.* Despite the changes that were brewing, Raye had seen and experienced enough.

"I couldn't see myself getting anywhere in Pine Bluff, unless I wanted to teach school," she said. "I needed to do something where I could work around people, but not in a learning situation, or in a negative situation. I felt like I had carried this load inside for too long, and I didn't know how much longer I could do it."

Raye was willing to give her young marriage a chance, and she had set aside enough money to pay her bills as she hunted for work. She knew she didn't want to teach, and she knew she couldn't stay in Arkansas. But she couldn't have imagined how her life would unfold next.

---

* One year after Raye left Arkansas, Melba Patillo was among the nine Black students who attempted to integrate Little Rock's Central High School, despite widespread backlash from White families. Rather than remain integrated, the school system shut down in 1958 in what was referred to as the Lost Year. One of the students, Gloria Ray Karlmark, left the state to finish her high school education and obtain a college degree, ultimately finding work in Europe as an editor and patent attorney.

II

A CAPITAL TIME

Raye Montague's US Navy portrait from when
she was deputy program manager of ships.

5

Exodus

Raye settled into her new life in Washington and looked for work, but she and Weldon were still navigating marriage problems that they couldn't seem to solve. Raye never discussed her issues with Weldon until much later in life. For years, when people asked her what went wrong, she would say the relationship "wasn't really there." In one video interview, when the questioner asked Raye about Weldon, she visibly grew uncomfortable and said, "I decided this was not what I expected in marriage." For decades, the story was not only that were they incompatible, but that Weldon was irresponsible.

"I never did know why they had trouble," his sister, Marge Coleman, said. "I just think they weren't well suited for each other. My brother was an introvert and Raye was not. Weldon was quite content to watch sports and read the newspapers, and Raye was very sociable. That's the only thing I can say. We really never discussed it."

What Raye and Marge did discuss was her need for a job. Raye and Weldon had purchased a duplex home in northeast Washington with his G.I. Bill. Raye's hope was that buying this residence would make Weldon settle down, keep a job, and behave more responsibly, but she claimed it did no such thing. According to Raye, Weldon had a hard time staying employed, and he preferred to spend his time hanging out with "the boys."

Raye and her first
husband, Weldon Means,
at Rock Creek Park in
Washington, DC.

In August 1956, Marge connected Raye with a recruiter for the Applied Mathematics Lab, who hired her as a clerk-typist after she convinced him that she knew a thing or two about UNIVAC computers she had never seen. For Raye, the job was a step toward working with these machines and becoming the engineer she always dreamed of being. For Weldon, Raye's work seemed to be an invitation to kick back and let his wife take care of things. By her telling, Raye had entered a period where she was trying to learn the ins and outs of a new job that was very important to her, and she was covering 75 percent of their shared expenses. It had become a lot to shoulder.

"I bluffed my way into a job working with people who had Ivy League degrees and who had worked on the Manhattan Project," Raye said. "I had never seen a computer, though I said I had. My job was to take the metallic tapes that others had typed data on and go on the UNIVAC computer, compare the tapes for errors, and see what changes needed to be made."

To Raye, the UNIVAC looked like the switchboard of an airplane with all its dials and knobs. But every morning she would go in to work, compare the tapes, and tell the computer operator, John McKenna, what fixes to make.

"He'd flip the switches, and the tapes would take off, and I'd make the corrections," Raye said. "Two weeks later, I asked him to teach me how to operate the computer, and he told me, 'No. If I did that, you'd have my job.'"

His words were prophetic. On October 20, 1957, Raye said, none of the engineers showed up for work to operate the computer. She waited for someone to arrive so she could compare the tapes. When no one did, she walked over to the computer, began flipping switches, and performed the daily type-ins. After weeks of watching McKenna operate the computer and eavesdropping on engineers, Raye had learned all she needed to know to get the job done.

"I'm the first person out there on the computer that day," Raye said. "It's humming. It's running. I'm mounting all the tapes, and I stood there. Nobody came out. I went over to the computer and I started to move switches."

The tapes took off, the printouts began, and coworkers started gathering around as Raye ran the computer. "I could see everybody looking at me," Raye said. "I have good peripheral vision. But I just started doing my thing. They don't believe we wear shoes in Arkansas or have teeth. How dare you touch our computer?"

Raye's coworkers knew she wasn't supposed to be operating the computer, so they went back and told her boss, Jack Smith.

"He came out and watched me, then walked over to where I was working and said, 'Raye, I didn't know you knew how to operate this computer,'" Raye said. "And I said 'I don't. But I know enough to get my job done.' And he said, 'Fine. You know more than any of the rest of them here, so from now on, this is your job.'"

She was promoted to computer systems operator. The next day, when the engineers came in to work, Smith told them to teach Raye how to operate the computer. After Raye became proficient with the UNIVAC, she was asked to teach other civilian nonengineers how to operate it too. Having theoretical knowledge and a fancy degree didn't matter so much anymore, at least in her case. After biding her time, Raye learned how to operate, code, and correct any programming issues she discovered. She was willing to work hard and get her hands dirty, just like she did when she was growing up in Arkansas. Armed with an education certificate, and extra UNIVAC classes she took on the side, Raye's attitude was not only that she could handle whatever came her way, but she could teach anyone else how to do what she knew how to do, too.

"I was training women and men, and I suddenly realized that the men were starting to earn more money than I was," Raye said. "I told Jack that I taught these men how to use the computer, and yet they were making more money than I was. So I asked him why that was. He told me it was because they had a car and could come in at night."

Because Raye didn't know how to drive, she used to carpool out to Carderock, Maryland, for work because the bus didn't run that far outside of the District.

"So I asked Jack whether I could make the same money as the men if I had a car and could come in to work at night," Raye said. "He said yes and thought he was done with it. But that evening, I went to a used car lot and bought a 1949 Pontiac for $375. I had the salesman drive it to my house, and I volunteered to work the midnight shift for a month."

At ten o'clock every night, Raye would slide behind the wheel of her Pontiac and begin the slow, careful drive out to Carderock from her house. Two hours later, she would arrive at the office, where she'd work until the day shift came in.

"When the day people arrived, I'd laugh with them and help them get set up for the day," Raye said. "They thought I was being kind, but I was waiting for the traffic to let up at nine-thirty, so I could drive home at no miles an hour. When I got home, I'd have to find a parking space I could just pull into because I didn't know how to park a car."

Around the time Raye was crawling to work and back in her car, one of her aunts died and Raye went home to Little Rock for the burial. Flossie had bought a car in the short time since Raye had been gone, and Raye drove it around town to practice.

"My mother had a friend who worked at the courthouse, and this friend said, 'Raye, I see you're driving your mother's car. Do you have a license?' And I said no. So he said, 'Do you want one?' And I said yes. A couple of hours later, he tossed it over to me, because at that time you didn't have to have a picture ID, and I didn't take an exam. When I got back to DC, the Department of Motor Vehicles honored my driver's permit and gave me DC permit. I proudly showed off my license at work, and everyone was laughing because nobody thought I had that kind of nerve."

Raye's boss came out to see what all the noise was about, and she told him she had a new driver's license. Confused, he said, "Well obviously you had one all this time." She told him she did not.

"He said, 'You mean you've been coming in here on this military base and driving and you didn't have a driver's license?'" Raye said. "And I said, 'Right.' So he told me, 'Well surely you had a learner's permit.' And I said no, because I didn't know how to drive. So he asked me who taught me how to drive and I told him I taught myself, driving in to work the midnight shift."

On the spot, she was promoted to digital computer systems operator and given a raise. She continued to work in the Applied Mathematics Lab, debugging higher-level computer programs for the UNIVAC 1 that were written in FORTRAN, COBOL, and ALGOL. There were other Black people working in the department, but all of them were mathematicians, not engineers. Although Raye could teach herself some of what she needed to know to do her job, she realized that she needed to improve her computer programming knowledge. She sent herself to a programming school offered by the Department of Agriculture and learned various types of computer coding skills.

"People thought I was weird," she said. "And I had this photographic memory, which blew them away. I could hold two conversations and listen to a third, and then do my own research. I guess they expected me to stay in the position I had when I started. But it was normal for me to grow and change. And I guess people didn't understand that."

As Raye grew and changed at work, she realized that the situation with Weldon had become untenable at home. In Washington, divorces took an extended period of time, so Raye first filed in Arkansas. Weldon agreed that she could have the house, but when it came time for him to sign the divorce papers, he refused. They shared the same house for two more years, after which Raye filed a separation complaint in Washington on May 16, 1962. In it, she acknowledged that she and Weldon were "reasonably happy for a very short time after the marriage." After that, the complaint states that Weldon "embarked upon a course calculated to and which actually did bedevil the plaintiff by his utter lack of marital responsibility, his refusal to remain steadily employed and by the actual infliction of cruelties upon the plaintiff."

At various turns, Weldon tried his hand at interior design and cab driving, but nothing seemed to last for long. Raye thought buying a house together would make Weldon behave more responsibly, but the complaint noted that she was paying most of their shared expenses. As a result, Raye was "reduced to a nervous wreck; her efficiency was impaired on the job and finally in order to save herself from a possible nervous break-down was compelled on or about the 11th day of March 1962 to remove herself from the marital abode."

The complaint said that while Weldon agreed not to harass Raye, he did just that, both at her home and job. On May 15, 1962, the complaint said Weldon "suddenly and without warning" forced himself into her car on a public street in the capital and "threatened the plaintiff with bodily harm; that while intruding himself in her car did pull the keys from the car while same was running and because of such conduct the plaintiff barely escaped being involved in a serious traffic accident; that the conduct of the defendant is wholly unprovoked on her part, and, is a further manifestation of his evil and cruel nature." Raye wanted quick action from the court, and asked for it to keep Weldon from bothering her, grant her a separation, and sort through their jointly owned property so it could be divided fairly. Finally, she asked for attorney and filing fees. Raye was granted a legal separation from Weldon on September 26, 1962.

"I was going through a lot of turmoil," Raye said. "I brought my mother up to live with me and we bought a bungalow on 3700 Thirtieth Place NE. She would cook and clean for me so I could work."

Although it is unclear what happened to Weldon at this time, certain documents indicate that he went back to live in Arkansas, possibly with family, and that he was incurring debts that were causing problems for Raye. In December 1962, Raye had an attorney write the local credit bureau to explain that Weldon's debts were not Raye's responsi-bility, and they were making it difficult for her to apply for credit.

Although Raye's marriage to Weldon was in the process of being legally dissolved, her friendship with his sister, Marge, remained. "She went her way, and he went his," Marge said. "But our friendship con-tinued. There was never any animosity as far as I was concerned. I always felt like she was part of our family."

6

Making Waves
in the Navy

Raye continued to work at the David Taylor Model Basin for the next eight years, steadily rising through the ranks into jobs with better pay and more responsibility. Her salary had more than doubled, from $3,155 to $6,435, since she was first hired as a clerk-typist, and her General Schedule, or GS ranking, had more than doubled, from a GS-3 to a GS-7. She was learning new technologies on the job as they were developed and training others how to use the new equipment that she was mastering.

In the early 1960s, UNIVAC developed a 115,000-pound computer that ran hydrodynamic simulations for nuclear weapon design. It was called a LARC, for Livermore Advanced Research Computer, and Raye oversaw its operation and maintained computer operator guidances for using it. She also became savvy in debugging and coding for it, seemingly increasing her confidence and authority in an office where she felt insignificant and ill prepared on her first day.

"I analyze problems arising during production and code-checking, and correct the routine when possible in order to complete a given run," she typed in an experience and qualifications statement dated October 30, 1963.

I conduct a training program to ensure that all operating personnel are thoroughly familiar with operating methods and procedures. I maintain liaison with Mathematical and Programming personnel and advise them of better utilization of the computer and assist them with finding errors in their routines. I check and proofread data and instructions prepared by programmers for computer runs. I prepare forms required for processing of work by other offices on the LARC computer. I give advice and assistance to programmers and mathematicians on matters affecting the use of the computer and auxiliary equipment. I maintain liaison with the management Data Processing Section with regard to their processing of LARC input and output data. I set up priorities on the work using various input and output medias in order to coordinate their schedules with the LARC operating schedule. I analyze computer problems and give the engineers sufficient details about machine malfunctions in order to allow the maintenance engineers to correct the source of trouble quickly and efficiently. I maintain a tape library for the LARC and I am responsible for keeping their maintenance files up to date. I provide recommendations concerning promotion or other recognition of personnel under my supervision. I program minor test routines.

Raye was in the thick of things and was the only person who knew how to operate the machine. Her boss, she said, knew nothing about the computer.

"I resented it, but that was the way it was at that time," she said. "People were bypassing my boss and coming straight to me to solve their computer issues. Eventually my boss resigned because he got sick of it."

Raye said she was asked to replace that boss on an interim basis. If everything worked out well, she was told she would be promoted two GS levels after six months. Raye decided to take her manager, Jack Smith, up on the offer. For the next half year, she ran three shift operations and taught people how to debug the LARC. When it came time to receive the promotion she was promised, Raye was told that it would not happen because there was a salary freeze. Raye said she looked around her and saw others getting promotions. Then she asked Smith was why there was a freeze for her but not for everybody else.

"And he said to me, 'I gotta be honest. You've got the right name, but you're the wrong sex,'" Raye recalled. "He said he couldn't stand women in supervisory positions, but they really needed me and hoped I would stay. He told me, 'I wish the guys worked as hard as you do.'"

Raye said she went back into the computer room and began filling out new job applications as obviously and quickly as she possibly could. Word got around the Applied Mathematics Lab that she was looking for a new opportunity, and the head of software systems development, Betty Holberton, approached Raye and asked if she'd come work for her. Holberton, an icon of early computing who had been instrumental in developing the UNIVAC, saw great potential in Raye, and she disapproved of the way Raye had been treated by her previous bosses. She planned to take Raye under her wing and give her a more supportive environment as they worked together on next-level computing for the government. The transition didn't go as smoothly as expected.

"At this point I had been working for eight years and the personnel people decided that I had to take the federal service exam," Raye said. "I took it and passed it with a ninety-something. They told me that my score wasn't high enough, and then they said the same thing after took it a second time and got another high score."

Raye was about to take the test a third time when Holberton told her to bring the exam to her when she was finished, and they would go together to take her results to personnel.

"I scored a ninety-five on this test," Raye said. "So when we showed it to personnel, they told Betty Holberton that they knew I was scoring high, but they had never had a non-mathematician or non-engineer working as a systems analyst or systems developer and they didn't want to establish a precedent for that now."

Frustrated, Betty Holberton asked the personnel staff to pull her own file. When they did, they saw that she had a journalism degree and had helped develop the ENIAC, which was used to study the feasibility of the hydrogen bomb.

Raye is seated front row at the dedication of the LARC Computing System at the Applied Mathematics Laboratory on May 15, 1961.

"Needless to say, I was transferred to Betty Holberton," Raye said. "And I worked for her as a software systems analyst and developer, taking whatever software was needed to operate the computer and modifying it so that it would be standard for the lab's operations."

In the meantime, Betty Holberton was so angry about the way Raye had been treated, she told Raye she would stay at the Model Basin until she could promote Raye from a GS-9 to a GS-11.

"Even though I had left Arkansas, you still had racism in Maryland and Washington, DC," Raye said. "Sometimes when I encountered it, it stunned me. By this time, I was working with White people at the Model Basin, and they accepted me for who I was: a human. I remember one day some of them asked me to go to lunch with them, so I said sure and jumped into the car with them."

The group headed to the department store Woodward & Lothrop, which had a nice dining room. "There were Black people with us, but they were lighter-skinned than I was, so the hostess looked at me and said to the group, 'We can't serve you in the dining room, but we'll take you to a private room and serve you in there.'" Raye said. "I have to give credit to the people who were with me, because they said if I couldn't eat with them, they weren't going to eat there."

Raye said her coworkers called all the people they knew and told them to stop shopping at Woodward & Lothrop, and if they had credit cards with them, to pay them off and get rid of them.

"The racism I encountered [in Maryland and Washington, DC,] was much more subtly done, but it happened," Raye said. "I just assumed in this particular case, that if you asked me to go to lunch with you, that it would be OK."

∞

Although the civil rights movement had begun to gain momentum while Raye was in college, she didn't join the ranks of those who marched or picketed, even though she did believe in the larger cause. She was a fervent supporter of President John F. Kennedy and believed that he would push the country to pass national civil rights legislation.

"I thought he had a lot to offer, and he had been on the naval ship, PT-109," she said, referring to the American patrol torpedo boat that had been rammed by a Japanese destroyer on August 1, 1943. Kennedy's actions to save the eleven crewmembers that survived made him a war hero and captured the imaginations of those who believed in his ability to unite and inspire an entire nation. Raye certainly had great hopes for him, especially with all the political, social, and scientific change that was happening in the country.

Not everyone shared Raye's optimism about Kennedy's commitment to the civil rights movement. Some believed that he was too tentative in his approach to helping Black people in the United States. As the son of Irish Catholic immigrants, he understood what it felt like to be discriminated against, and he felt there was a moral imperative to treat people equally and fairly. However, he was a foreign policy aficionado at heart, and it would take him some time to understand the full extent of the discrimination Black Americans faced. Plus, he believed there would be political repercussions if he came out solidly in support of equal rights and opportunities for Black people. So Kennedy walked a tightrope between wanting the support of Black voters and wanting the support of southern Democrats who embraced segregation. The Reverend Martin Luther King Jr. was growing increasingly frustrated with Kennedy's desire for order at the expense of moral progress, and he organized a demonstration in Washington to push for a strong federal civil rights bill. It had been a long time coming.

On Wednesday, August 28, 1963, Raye made her way down to the Lincoln Memorial Reflecting Pool to hear Reverend King deliver his "I Have a Dream" speech to the more than two hundred thousand people who had gathered there.

"I was living his dream," Raye said. "And to hear him speak was marvelous." It was an exciting moment that inspired Raye and the countless others who had filled in around her.

"They brought people in from all over the country on buses," she said. "And when it was over, everyone quietly left. There was no free-for-all, nothing. But nobody expected any fighting or anything after something like this."

After the speech, King met with Kennedy and his vice president, Lyndon Baines Johnson, to stress the need for bipartisan support of comprehensive federal civil rights legislation. Such a meeting would have been unthinkable a decade before, at a time when Raye wasn't allowed to study engineering because of her color. The promise of what could come next was inspiring, and Raye looked to both these men as beacons for what should be. She was a government employee, but that fact didn't mean she felt like she couldn't protest government policy. As a matter of fact, the fact that she was a federal employee made her well placed to protest in her own way and on her own terms.

"My mom believed in civil rights for everyone," David said. "She didn't care about your looks, your orientation, your religion. She believed everybody had the same rights, and she would support you as an individual if you wanted to make a change. But she would also let you know that you had the obligation to do the same for other people."

David said his mother later understood the implication of her being on some of the boards she was on. Not only was it a good opportunity for her, it was an inspiration to people like her, and it sent a message to those who were uncomfortable with her being there.

Whether it was civil rights, or technological advances, the nation was undergoing big, ambitious changes. Black and White youths traveled through several southern states to expose unlawful segregation in interstate bus travel. After Alan Shepard became the first American in space, President Kennedy promised a man on the moon by the end of the decade. James Meredith became the first Black student at the University of Mississippi, and in the following year the Southern Christian Leadership Conference began a movement to show what happened to Black Americans who attempted to integrate public spaces in Birmingham, Alabama.

Not all of those changes were welcome. Even those that were didn't happen overnight. But there was the sense, however fleeting, that there could be something good on the horizon, and that things might be better, someday. Hope bloomed like the soft pink cherry blossoms that surrounded the capital each spring. However, like the beautiful flowers, this optimism would be short-lived. On Friday, November 22, 1963, Raye said she was running through the halls at the David Taylor Model Basin happily telling coworkers about a colleague's newborn.

"That's when we found out that President Kennedy had been assassinated," she said. "We just stopped everything and left work."

The nation was glued to television sets that entire weekend, watching as *Air Force One* arrived in Washington, DC, that Friday evening with Kennedy's body on board. Lyndon Baines Johnson was now president. Martin Luther King Jr. said he was "shocked and grief-stricken" about the assassination. "The finest tribute that the American people can pay to the late President Kennedy is to implement the progressive policies that he sought to initiate in domestic and foreign relations," King said.

But first, there was mourning. The world grieved as it watched four full days of television, much of it commercial-free. Kennedy's body was brought to the Capitol Rotunda so that Americans could pay their respects. Raye and Flossie were sitting in front of their set when nightclub owner Jack Ruby shot the alleged assassin, Lee Harvey Oswald, in the basement of Dallas Police Headquarters. Raye wondered what was happening to the country, and immediately changed the channel to watch mourners file past President Kennedy's flag-draped coffin at the Capitol.

"I said to my mother, 'You know if we lived any other place in the world, we'd be saying that we'd be at the Capitol if we lived in DC. So come on, let's go,'" Raye said. "And we went down and stood in that line for eighteen hours. The line wound all the way around the Mary McLeod Bethune monument that's there in Lincoln Park, but we waited and were able to pay our respects. My aunt in Little Rock said she saw me and my mother go through the line at four o'clock in the morning."

Raye said she had such great hopes for Kennedy and his vision for the country. It seemed like just yesterday that Martin Luther King had led the March on Washington for Jobs and Freedom, and she said she was keen to see how Kennedy's and King's vision for civil rights could push the country forward. Despite her grief and shock, she said she knew that she had exceeded many people's expectations for her. This moment only strengthened her resolve to keep pushing, to keep fighting.*

No matter Raye's determination, everything seemed like an uphill battle. She was going through a divorce and fighting for herself at work, but she didn't want to let anyone except for her mother know that she was feeling vulnerable, or even why. She needed people to see that she was strong, charismatic, and capable of doing anything. If she felt threatened or hurt by someone, she would never let them know that they had gotten to her. There was no sense in giving them that sort of power.

And then a coworker introduced Raye to a handsome barber named Dave Montague.

Raye said she was always so busy working she never had time for dating. Men were interested in her, she said, but they would often drift

---

* Less than a year later, President Johnson advanced Kennedy and King's vision by signing into law the Civil Rights Act of 1964, which outlawed discriminatory practices in voting, hiring and education. It also provided for integration of public places such as restaurants and schools. In 1965, he then signed the Voting Rights Act, which gave Black voters the legal means to challenge any voting restrictions they faced on the local level. Decades later, in 1995, Raye's son would become the senior investigator for the Kennedy Assassination Records Review Board, which was a source of great pride for her.

away after a period of time. Raye chalked it up to their insecurities, but when she met Montague, things were different. He was smooth, handsome, and seemingly very self-assured.

"He was a tall, dark chocolate yum-yum," Raye said. "Oh, he was a hunk. Smart, danced beautifully, played chess, played bridge, you name it."

It was 1964, and Raye's divorce to Weldon was not yet final. After how things had fallen apart with Weldon, Raye was frankly hesitant about falling in love. But Dave was impossible to resist. His looks and charm aside, he was a successful barbershop owner who once appeared on *The Tonight Show with Johnny Carson*. According to Raye, *The Tonight Show* filmed an ongoing segment on talented workers across the country, and Dave was invited to appear to talk about his flair for doing hair. Dave did well for himself, and Raye joked that because he was a womanizer, he likely used his appearance on Carson to meet and impress more women. As it stood, he had no shortage of female admirers who would have loved to take him off the market. These were the early signs that he was a catch, and also that Raye would find herself caught.

Over time, though, troubling details about Dave emerged: He was the absentee father of a preteen girl named Debra, who was being raised by her grandparents. He had drug and alcohol problems, and a prison record. There was also a string of women he was unwilling to ignore. Yet he and Raye fell in love and married on October 15, 1965, less than a month after her divorce to Weldon was finalized.

Within months, Raye was pregnant and suffering through two hours of morning sickness every day before working an eight-hour evening shift. She spent her nights testing timesharing, a process in which multiple users could perform different functions on a computer at the same time, thus increasing its efficiency.

Prior to this, when computer users entered information, there was typically some wait time for a response to that action. As computer memory and speeds increased, it became possible for more than one user to work on the machines. With timesharing, the computer could be programmed to remember where one user was during their process, and start performing functions for another user or users. As time-

Raye with Dave Montague on their wedding day.

sharing was being developed, David Taylor Model Basin was one of the beta sites, and they needed someone to test the system from 4:00 to 6:00 PM. These hours were convenient for Raye's new schedule, so she would work with the companies who had developed the program and the employees who were creating ship models using the numerical control tapes. Once she was done troubleshooting, she advised the system people about the types of problems people were running into so they could make changes. It was exacting work—made even more exhausting by her pregnancy and a husband who could not seem to turn a blind eye to other women.

"One of the things that hurt me, I remember, was that he kept saying to me, 'Hard as you work, you should be promoted,'" Raye said. "I said I agreed. Well, I was pregnant, wearing my maternity clothes, and I had a promotion that I showed to him. I told him, 'Let's go out and celebrate,' and his whole confidence changed just like that. He told me, 'That's how the White man keeps his foot in the Black man's ass, by promoting you women.'"

Raye was stunned.

"After that, it just sort of went downhill," she said. "I think that some of the guys he was hanging out with were talking about women making more money, doing things, or excelling and maybe he didn't mean to hurt me in that fashion, but he did. It really hurt."

On the other hand, she and Dave were about to become parents, so she was likely hopeful that they could work things out. Raye applied for and received three and a half months of maternity leave, which she began on August 1, 1966. Nine days later, her son, David Ray Montague, was born.

Whatever problems Dave and Raye had prior to the birth seemed to fall by the wayside, at least temporarily. The birth announcements they sent to family and friends exude sheer parental delight:

David Ray Montague
is my name
Wednesday, August 10th, 1966
is the day I came
4 lbs. is what I weigh
and these are the folks
with whom I stay:
Raye and David Montague
3700 Thirtieth Pl. NE
Washington, DC

David was a delicate baby, likely because of his mother's smoking habit. Given the tentative place Dave and Raye were in their marriage, it is easy to see how the initial joy of having a newborn could be replaced with the anxieties and challenges of nurturing someone so frail. As a good-time guy, Dave, who had previously walked out on his daughter Debra, was likely thinking about how to make a quick exit from his taxing new reality. True to form, he left when David was nine weeks old, leaving Raye to care for him on her own. She had less than two months to figure out how she would care for David once she went back to work.

As far as Raye was concerned, it was time to kick like the devil and holler for help. Flossie had moved back to Arkansas once she was

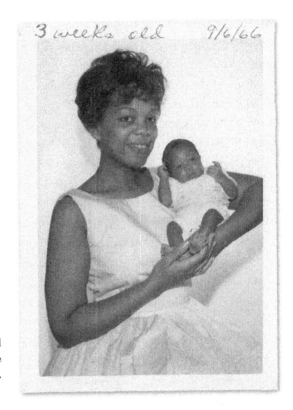

3 weeks old   9/6/66

Raye with David Montague at three weeks old.

confident that Raye had gotten back on her feet after Weldon. But this was an entirely different situation. Raye called her mother and told her she needed her. Flossie didn't hesitate; she sold her salon in Pine Bluff and moved up to Washington permanently to help her only child and grandson. After all, a mother's unconditional love and encouragement have a way of steadying an uncertain child, no matter their age or circumstance.

Flossie moved one of her two salon chairs and a heavy hair-dryer into Raye's house, perhaps as a reminder of the profession she embraced that enabled her to take care of herself and daughter after her divorce. Her things in place, she took to looking after David, cooking, and cleaning. That way, when Raye went back to work, she could focus on her job. Then Raye's father, Rayford Jordan, reached out to her for the first time since he and Flossie divorced. Flossie had sent him the birth announcement, and he wanted to see his only grandson.

"David was the only grandchild he had because my father never remarried," Raye said. "I asked mother if it was OK with her, and she said it was."

Rayford Jordan came to Washington to stay with his daughter and ex-wife for two weeks. During his visit, he reflected on the family's early years together and how his struggles with alcoholism made it difficult for him to be a good provider. Now, he watched as his thirty-one-year-old daughter and namesake, Raye, was supporting both an infant and Flossie, and he felt that it didn't need to be that way.

Rayford wanted a chance to make things right. He returned to Mississippi and considered the twenty-six-acre portion of the family farm he had been given. The land was not to be divided or sold until grandkids came along. Although he and Flossie had long been divorced, and though he had not seen Raye since she was four years old, David was still his grandchild. The condition of his inheritance therefore being met, Rayford pled his case to divide his portion and succeeded. He then deeded the land to Raye. Rayford told her she could rent it to someone to live on or sell it. The easiest thing for her to

Raye, and her mother Flossie McNeel (seated at the head of the table) with David on her lap, Thanksgiving 1966.

do would be to have the timber on the property thinned out and sold from time to time so she could keep a steady stream of cash coming in. Whatever she decided, Rayford wanted her to take the land and do what was right for her and her little family. Knowing the parcel was hers and that her mother was home with David was a great comfort to Raye as she headed back to work.

"I was a single parent and fighting the war at work," Raye said. "My mother was living with me and could help me with the baby while I worked. My father gave me this land. Betty Holberton then wrote me up for a raise, telling me I was well past due for it. Then she left, but not before giving me a copy of that promotion 'in case they decided they lost it.' Betty set up the same kind of operation at the National Bureau of Standards, but didn't ask me to come with her. She said I had a role to play where I was and I agreed."

Raye said Betty was a godsend to her. "Everything that happened to me, happened for a reason," she said. "I had to face all these obstacles to be ready for the big thing that was in store for me. I had to go through all the traps, hard work, and frustrations in order to be ready for whatever it was I was sent to do."

7

A Change Is Gonna Come

Once Raye returned from maternity leave, she was back on the computer, teaching people how to use the timesharing system and biding her time for something better. She was frustrated and felt unsettled without Betty Holberton there to defend and protect her at work. It did not help that she had commute through more than forty miles of Washington traffic each day. She was also mired in an on-again, off-again relationship with her husband, who was a barely there parent to their young son.

"He didn't really have much for David once he was born," Raye said of Dave. "Oh, he would come by once in a while. David loved him at first. He just loved him. He would stand at the window and wait when he'd call and say he was coming. And then, when Dave wouldn't show up, David would have an asthma attack. I ended up telling Dave, 'Don't tell the child you're coming. You call me and see if it's OK, and if it's OK, fine, you can come by. But I'm not gonna tell him. I'll just let you show up."

For a while, that arrangement worked. Then, David was diagnosed with chronic asthma. Raye struggled to keep up with running him to the hospital and to various doctors—and paying those bills. Dave

David Montague in
one of the chairs at
his father's barber-
shop, date unknown.

stopped coming with any sort of frequency or providing any sort of
child support. As far as Raye was concerned, he had burned a bridge
with her.

"I never told David all these things," Raye said. "I would never
speak bad against his father to him." She easily could have. One night,
Dave came back and stole her car so he could meet his buddies for a few
games of cards. When Raye realized her car was gone, she automati-
cally knew who the culprit was, and where he might be. She recruited
a neighbor, they armed themselves with pistols, and then they drove
past Dave's haunts in search of her automobile.

"As the story goes, they found the car, and they stole it back,"
David said. "Mom's attitude was that nobody was going to steal from
her, not without a fight on their hands."

David said that in retrospect, he knew his mother was in a hard
place. She still had feelings for Dave, but she also knew that he was
bad for her. Yet, she wanted Dave to have a relationship with their son
under safe, controlled circumstances, but David was becoming indif-
ferent to him. Raye was doing what she could to give David everything—happiness, structure, a life she never had—but it was no effort-
less dance. In later years, David told his mother he remembered Dave
coming around to visit her. To this day, he has an image of them sitting
alone in her bedroom, talking, but Raye told him he couldn't possibly
have remembered that. It's possible that Raye and Dave struggled to

**Raye with David on
his third birthday.**

let each other go, and as strong as Raye tried to be, she held on to hope
that they would reconcile. Telling David he couldn't possibly have
remembered her speaking to Dave in her bedroom could have been
her way of telling herself that once Dave walked out, she never would
have let him into one of her most personal spaces. David said his recollection wasn't about a little boy being wistful that his parents would
reunite. It was about being quietly observant, as his mother once was,
and stating a fact. No matter the confusion about their marital status,
Raye marked the day that Dave Montague walked out on her and their
newborn son as the day she became a single parent.

"There was a lot of love, hurt, and tangled emotions with him,"
David said, adding that his mother taught him a very important lesson
about getting closure with certain relationships. "With me, once that
door's closed, it's closed."

With his mother, it was a different story, at least with Dave.

⊶∞⊷

On April 4, 1968, Raye left work late and inched through traffic after
learning that Martin Luther King Jr. had been shot at the Lorraine
Motel in Memphis, Tennessee. As she reached the district line in her
car, she said military officers stopped her. They weren't letting people

anywhere near the city. David was almost two years old at the time, and although he was likely fine at home with his grandmother, Raye wanted to get home to him, especially on a day like this.

"They told me, 'Lady you can't go in there,'" she recalled. "And I told them I lived just two blocks inside the district line. I had to get in there, I told them, because my baby was in there. And then they told me that if I went in there, I couldn't come back out. So I said, 'I don't care about coming out, I just gotta get in there to my baby.'"

They let her in.

Raye didn't see any unrest in her neighborhood, because she lived in the northeast part of the city, and the looting, arson, and clashes with police and firefighters were happening downtown. She only saw what was broadcast on television.

"I woke up, I think the next day, and there in my front yard was a window air conditioner," she said. "Somebody had stolen it and dropped it there."

She couldn't raise her son in an environment like that, so she put her house on the market, and it sold before she could find a place to move.* She found a home four blocks across the district line in Hyattsville, Maryland, where she enrolled David in a Montessori school. While it made her feel safer, it made her commute even harder.

"Mom wanted to move somewhere where riots wouldn't be an issue, so she found West Hyattsville, which was a series of residential neighborhoods," David said. "She liked that it was a mixed community, and that there were no vacant houses, no failed businesses, and that kind of thing. It was a nice community and she felt good about us being there because it was a real neighborhood."

David said his mother used to correct him when he told people he lived in Hyattsville, because at the time that was thought to be nothing but industrial parks and business complexes.

"It is *West* Hyattsville," she would tell him, and it took him a while to understand why she stressed that. To her, there was a difference,

---

* One week after Dr. King was killed, Johnson signed into law the Civil Rights Act of 1968, which went a step further than the similarly named act signed in 1965 because it prohibited discrimination concerning the sale, rental, or financing of housing.

and that difference was the safety and stability she could not get in the city.

"She said when we first moved into the neighborhood, there was a man who gave her hell for years because he thought she was an unmarried woman moving into a family neighborhood," David said. "It was always a mixed neighborhood, but this man thought it would bring in a bad element to have a single parent there; that it would be party central at our house."

The neighbor's concerns were not rooted in race, David said, because the man in question was Black too. "He just didn't trust anybody," David said. "He was always very particular and wanted things a certain way. He would pick at you if you parked too close on the street in front of his house; he'd pull his own car two centimeters away from your bumper. He just had a really negative view of everyone. Once he got to know her, they became buddy-buddy. It took a while for him to realize that my mother was OK. My mom taught me that it's not always the person, it's about how they were raised. Sometimes people can't see you as an equal right away."

As Raye and her family established themselves in a new neighborhood, there was still the issue of her unresolved relationship with Dave Montague. Although Raye said that she and Dave were finished nine weeks after their son David was born, it is evident that they remained married for the first years of the boy's life, even though they may not have always lived under the same roof.

David recognized that his father was not making an emotional or financial contribution to his life. He asked his mother about it, and she explained that she was the one paying for the house, the car, the clothes he wore, and his father wasn't contributing anything.

"David said to me, 'Couldn't he give you any money for me?' and I said, 'Yeah,'" Raye recalled. "Then David asked, 'Well why doesn't he?, and I told him, 'That's a mystery.' But I also told him that I saw to it that he didn't want for anything."

David went upstairs to his room for a half-hour, then came back downstairs to talk to her. He told his mother he wouldn't call his father Daddy anymore.

"I said 'Why is that?'" Raye recalled. "David told me, 'Because he doesn't do what a daddy is supposed to do.' I told him that was between him and his father, but I was his mother until the day I died and not to forget that."

David started calling his father Dave. Dave never questioned it, and Raye never told him why, but David refused to call him Daddy for the rest of his life.

"Mom knew I resented Dave, and that I saw him as the opposite of what I wanted to be in life," David said. "She always told me how important it would be for me to stay out of trouble, and she had her male friends talk to me about how to survive as a Black man in this world. They told me how to handle traffic stops and said I needed to keep my hands outside the car window if a policeman ever stopped me. They told me to always carry identification with me because people would always assume I didn't belong there. They told me to always look people directly in the eye, to not give people more information than they needed about me, and to think twice about the consequences of every decision I made."

Raye didn't want her son to be a statistic, and she was grateful to have a support network that could provide David with the extra guidance. When David was younger, she enrolled him in karate classes after hearing that he was getting picked on by other kids.

"I remember one day when I was in grade school, some bully was trying to mess with me on our front lawn," David said. "I hurt his arm, and when Mom found out about it, she said, 'Well I guess those karate lessons are money well spent.'"

Advice and self-defense classes aside, David said he still got stopped and harassed while he was growing up, whether he was outside a convenience store eating ice cream, walking home from school, or strolling down the aisles of a store. He would be followed, told he didn't belong, asked to leave. At least he knew how to handle these situations in a manner that kept him safe and out of trouble. Dave wouldn't have been able to offer him any wisdom that would have kept him protected.

Dave's relationship with Debra Moore-Lewis, his daughter from a previous marriage, was even more fraught. Although Raye and David knew of Debra's existence, she was not a part of their lives, and it's difficult to know whether it was because Dave wanted to compart-mentalize certain parts of his life, or whether Debra preferred not to be in his company. Whatever the case may be, Debra said she did not meet Raye until well after David was born. Dave was no longer living under Raye's roof, but a mutual friend of theirs had died and he had Debra drive him to the funeral home.

"Raye came out to the car and insisted on meeting me," Moore-Lewis said. "My father never introduced her to his own mother either. She had to insist on meeting her too. But Raye? As I got to know her, she accepted me and made my heart melt because I felt like she loved me more than my own parents did. I could talk to her like a daughter should be able to talk to a mother and not feel ashamed or judged. She told me she loved me and that I was the daughter she never got to have. At this point, I hadn't met David in person, but I knew of him because he used to call to talk to my dad on the phone. He and my father didn't have that great of a relationship, just like I didn't. My father wasn't easy to get along with. But Raye Montague? She embraced me, even with all the dog stuff my father did. She never held that against me. That's a real woman right there."

Considering Dave's impact on her and his own children, one wonders why Raye didn't file for divorce sooner. After all, she said that women in her family believed in getting out of a relationship if it wasn't working. Perhaps she was too overwhelmed to file for divorce in the midst of working a demanding job and raising a child who strug-gled with his health. But after five years of marriage, during which Raye knew Dave continued to see other women and contributed next to nothing to David's life or hers, Raye decided she had finally had enough. When she told Dave that they were through, for real this time, he told her that he would never make it without her.

In a letter dated September 30, 1970, attorney Karl M. Dollak notified Dave that Raye had come to his office seeking a divorce. Dol-lak noted because Raye hadn't heard from her estranged husband in several months; there was no reason for her not to file for divorce as they had been voluntarily separated for some time.

"Your wife advised me that she felt it necessary, primarily for rea-sons of her health (she tells me that she advised you that she is going into the hospital this Thursday for surgery, and will be away from her work for six weeks) and she also has advised me that you have agreed that she proceed to get the divorce with no cost to you," Dollak wrote.*

Dollak said he had already begun the divorce paperwork, and that if David was in favor of proceeding, he'd arrange to have him represented by an attorney at no cost. Dave had fallen on hard times and was in rehab in Kentucky, so as complicated as their relationship was, Raye didn't want to make a hard situation even harder for him.

"I understand that you have apparently had more than your share of troubles since I last saw you, and I hope the future is brighter, and you have my best wishes," Dollak concluded.

When the divorce was final, Raye turned to her family for solace, only to become a target of her aunt Angeline. "She told me if I didn't watch it, I'd wind up like my aunt Gladys, who had been married a lot," Raye said. "And I told her that the Bible entitled me to seven husbands and I intended to get them all. No one else would speak up to her like that."

Something had to look up for Raye and soon. One day in 1970, Raye said Gilbert Gray, the head of the Applied Mathematics Lab, came into her office and ordered her to leave immediately for a meeting at the main navy offices. Raye said she grabbed her purse and headed out to the shuttle so she could make the meeting on time.

At the time, Raye might have felt tired and thwarted at every turn. But over the years, she had impressed someone who saw a role for her in a nascent department called Computer Assisted Ship Design and Construction, or CASDAC for short. When engineers realized that computers could be programmed to run machines, they envisioned using the process to design and manufacture products. The navy wanted to determine what benefits they could get out of using computers to design ships and submarines. At the time, they built billions of dollars' worth of ships and submarines each year, using thousands of workers, from highly skilled professionals to laborers. The highly

---

* David Montague said his mother was having a hysterectomy at that time, and a letter from an obstetrician during that period confirms that she would need to miss work for six weeks. David said that a hysterectomy was presented as her only option, when there were other methods of birth control available. Raye was upset because she had hoped to have another child and those other options were not explored.

skilled workers were not only involved in the entire ship acquisition and maintenance process, they were also responsible for determining what products would be developed and manufactured for the navy.

In 1966, the US Navy had begun the program called CASDAC to investigate and analyze how technology could be applied to the entire shipbuilding process. CASDAC was the brainchild of Wallace "Wally" Dietrich, a civilian engineer with a degree from Johns Hopkins University who held several patents for ship-design technologies that he dreamed up in his spare time. He had been an employee of the navy since 1940, methodically working his way from the Bureau of Ships to the Naval Ship Engineering Center, and then to the Sea Systems Command. By all accounts, Wally was a soft-spoken and courteous man who rarely got angry. Some former colleagues said they don't believe he ever swore. He was widely considered to be a knowledgeable visionary with an innate understanding of where his profession was headed. His foresight and his flair for nurturing people made him a natural fit to lead an office of specialists in various aspects of shipbuilding and computer sciences.

Dietrich's crew worked out of the Naval Ship Engineering Center (NAVSEC) in Hyattsville, Maryland. NAVSEC was an ideal location for CASDAC because of its responsibilities for the design, engineering, and support of the US Fleet. Overall, two thousand employees worked at NAVSEC, fifteen hundred of whom were engineers and naval architects.

Although CASDAC had a small team, it was a crucial project as far as the navy was concerned. After World War II, it had begun looking for new ways to replenish its dated ships. Tensions with the Soviet Union were rising, and the United States needed a quick solution for what was becoming a very heated arms race.

For centuries, shipbuilding had been a long, laborious process. After an admiral ordered a new ship design, a naval architect would take the dimensions that had been given and begin sketching a hull— or main body—that might fit the request. Such a draft would take hours, much of it spent consulting specifications and mathematical equations that needed to be considered in the overall design, things such as length, water displacement, and so on. After an engineer had completed the first draft, there would still be successive drafts to create, each of them showing different design options for the same requirement. All of them had to be carefully calculated to meet an admiral's initial request right down to the pipes, wiring, weaponry, materials, and other particulars.

Any alteration—big or small—on a boat that had limited space to begin with would require vast amounts of reconfiguring, redesigning, redrafting, and recalculating by a lot of different people. If an admiral decided the ship needed to have room for two extra people beyond what he had originally anticipated, a design would have to be adjusted to accommodate these sailors, who would need to be factored into the air conditioning, plumbing, and space for sleeping quarters. Extra amenities led to extra weight, which would require the ship to have extra length and propulsion. Once these extras were added, designers were often forced to make other adjustments to the existing draft.

The changes were not always straightforward. Moving a switchbox over six inches could put it on top of a steam pipe, which if it were moved, could interfere with the entire ventilation system. Manipulating these alterations so that the engineering remained sound, and the systems integrated, was something of an art, and a painstaking one at that. Forget how one manipulation affected something else, or make one mathematical mistake, and it could cost the navy millions of dollars in fixes down the road. Designers needed to get things right before the ship went from a draft to a blueprint, which is why the navy began looking for ways to prevent the chaos that big changes—or even slight ones—wrought.

Considering the size of the average navy ship and the number of systems it includes, it would be no small task for a human to know and remember all of the variables in a way that would make any design change easy, or swift. No person has a memory that large, and no person would be able to solve the thousands of mathematical problems that would arise from shifting one little dial just a little bit to the left without burning themselves out. Any human who did this would likely have to consult multiple blueprints and experts to get the job done. Such endeavors took time. And of course, time was money.

But if you needed a good memory and the ability to perform countless complicated calculations, there was no better solution than a computer. Theoretically, it could store the necessary details of each ship system, and notify engineers of problems before the process reached the construction phase.

The US Navy began experimenting with the use of computers in its ship design and construction process in 1948. By 1970, when Raye Montague was under orders to rush downtown for a meeting that would ultimately change her life, the navy had more than 165 ship computer programs in operation, and 40 more under development. Computers were used for everything from conceptual design modeling of ships, to the development of detail design and working plans for construction. Although the CASDAC operation at this point was small, most computer installations were massive affairs that sometimes took up three floors of an office building. CASDAC employees had to send some of the programs they wrote out to the David Taylor Model Basin to run them on the large, memory-intensive computers there. Many times, they'd request Raye Montague's help with their work.

One of those employees was Art Fuller, a mathematician from Birmingham, Alabama, who had been hired to develop some of the equations for computerized ship designs. He was the only Black person in CASDAC, and sometimes felt it, even though Wally Dietrich did everything he could to make sure Art felt at ease among the predominantly White staff. According to Art, Wally was well aware that he had to walk through a room of forty White naval architects on the way to his office. Dietrich was protective, and although there were no racial incidents, he moved Art's desk closer to his just to keep it that way.

Because Art had been working on a top-secret nuclear submarine project, he had to load two program boxes into a chauffeured car and take his work out to David Taylor personally so he could sit in the room while Raye ran his program and then wiped the computer clean so the technology wouldn't be exposed. It didn't take long before Art decided it was pretty unfair for a woman with a bachelor's degree to be stuck in an underling role.

"First of all, she looked like me," Art said. "Second, she was more

Raye portrait, date unknown.

than competent and she had a bachelor's degree. So it just didn't sit well with me that she was being treated as a subordinate."

Art never forgot about Raye, and he recommended her for a computer operator role within the division. Raye and her bosses at

the Model Basin didn't know that this was the meeting—a job interview—she needed to get to, and fast.

"By the time I got downtown, there was a guy there named Dick
Hanley who wanted to talk to me," Raye said. "He said they had heard
a lot about me and wanted me to come work with them. He asked me
what it would take to get me to come and I said a GS-12 to come and
a GS-13 within a year."

Hanley told Raye he'd have to talk to management, and she got
back on the shuttle for Carderock, where she likely started to rumi-
nate on the benefits of such a move. In an undated letter she originally
addressed to "Mr. W. Dietrich" before crossing it out and writing "To
whom it may concern" instead, she noted the reasons why she would
like to transfer to CASDAC. The first reason was that she was com-
muting twenty-one miles one way from her home in Hyattsville, and
NAVSEC was two miles from her house. She tried to make the second
reason as compelling as possible, noting that her son went to school
one-half mile from NAVSEC. She began to write, "it would improve
his," before crossing it out and writing "it would give our family," and
crossing it out again. Then, she resolved that working at NAVSEC
headquarters "would make his schedule more realistic if I were in the
general area."

Looking back, David said his mother went out of her way to spend
as much quality time with him as possible. Having a longer commute
in DC rush-hour traffic complicated that desire, so a job at NAVSEC
was a way out of a hard situation for a woman who wanted work-life
balance.

But her letter was not just about what *she* needed. Raye wrote that
CASDAC needed someone who did the kind of work she performed
in Carderock. "The computer-related duties, accounting analysis, and
consultation are similar to the duties I desire to perform at NAVSEC,"
she wrote, before crossing it out and scribbling, "I would like to assist in
the planning, development, and implementation of computer administrative procedures for NAVSEC computer facilities and services."

Raye never finished the handwritten draft of the letter, and it
is unknown whether she ever typed another version and sent it to
NAVSEC. It's clear that she saw an opportunity that would be good

for her not only professionally but personally. When Raye returned to her office, she said her boss, Gil Gray, yelled at her for leaving a dirty coffee cup on her desk.

"I tried to tell him he told me to leave right then and there for that meeting and that I didn't have time to put it away," Raye said. "While he was jumping all over me, the phone rang and it was Dick Hanley. He told me they would honor my GS requests, but I had to come in two weeks, or else they would lose the funding. I said I'd take it."

Raye said she didn't tell Gray about her pending move because other people had been up for good jobs in good places and he had blocked their transfers. "I knew that if he knew I was up for this job, he'd block me too," Raye said. "So I went to the administrative assistant for the Applied Mathematics Lab and told her that when word comes to release me to Dick Hanley, release me, because otherwise Gil will block it. She signed off on it."

When Gray heard that Raye was leaving, and that there would be a going-away party for her, he approached her desk and asked her where she thought she was trying to go.

"I told him I was leaving that Friday," Raye said. "He said I couldn't do that because he had all these other things he needed me to do. Well, I went to the going-away party, and I went to personnel and turned in my badge, and I signed out on annual leave for the next day. You couldn't stop me."

8

Impossible Tasks

When Raye reported for duty as a systems analyst at the Naval Ship Engineering Center, she knew that she was walking into an experimental program that Congress was not eager to fund. In 1968, President Lyndon Baines Johnson was on the verge of going before Congress to advocate for foreign ship construction. He believed the high cost of American-built ships was the main obstacle for expanding the naval fleet. It was an election year, however, and Johnson rethought his position because he feared backlash from the labor movement.

Richard M. Nixon won the presidency that year, defeating Johnson's vice president, Hubert Humphrey. Where Johnson wanted to outsource the nation's shipbuilding, which had become costly for the navy as well as for commercial shipwrights, Nixon wanted to keep the industry within US borders. He proposed to build three hundred ships over the next decade, offering subsidies to shipyards to make sure the quota was met and that no buyer turned to a foreign builder to construct their fleet. Under the law at that time, the United States funded up to 55 percent of the cost of commercial ship construction, which represented the difference between the cost of building ships domestically versus overseas. Under Nixon, the subsidy would gradually drop to 35 percent by 1976. Faced with the challenges of keeping the industry in the country, the conditions were ripe for some engineer to develop a way of building ships faster and more economically. The question was what that way would be.

The navy believed CASDAC might be the answer. It could produce the technology that would help it keep pace with the growing Soviet fleet, as well as keep shipbuilding in the United States. To Raye, this was an exciting opportunity to be in on the ground level of something new that would allow her to experiment and maybe even have a hand in something important. Best of all, her new boss, Wally Dietrich, was said to be a gentlehearted leader who encouraged his charges to think for themselves and find creative solutions to the problems they were trying to solve.

Given the challenges Raye faced as a Black single mother who often needed to work harder to prove herself, though, it's likely that she walked into a new job in a new environment expecting to fight with a new boss. After all, she had battled her way out of Arkansas, through the David Taylor Model Basin and over to CASDAC, so why would this be any different from anything she had encountered before? She recalled the day she met Wally; he was coming in to meet "the new guy." Raye never used a title such as "Ms." or "Mrs." with her own name, so it is likely that Wally assumed that Raye was a White male and was surprised to meet a Black female instead.

"I extended my hand and said, 'Hi. I'm Raye Montague,' and I kept my hand extended and stared him in the eye," Raye said. "When you do that, someone has to shake your hand, or else they'll be embarrassed if someone else is around. So Wally said, 'You're Raye Montague?' He finally shook hands with me, and his attitude seemed to be, 'What do I have to do to get rid of her?'"

According to Art Fuller, Raye's race wouldn't have been an issue for Wally. Yet Fuller acknowledged that if you had been subject to racist experiences before, as he and Raye certainly had growing up in the South during Jim Crow, you would expect to be mistreated by a White person with any degree of authority.

"When you have experiences like the ones I had grown accustomed to in Alabama, you don't forget about them," Art said. "They're there in your mind when you're walking through a room of people who are just doing their job. But how you're interpreting the experience, based on what you know, is different. It's not hard to see how Raye, especially being a woman, would have these challenges, too."

In fact, Fuller said, Raye's internal dismay was likely compounded because she was both Black and a woman. Whatever the case, she was walking into an all-male environment where there was intense pressure to advance the art of shipbuilding. Under the circumstances, Raye felt a profound need to prove herself.

The US Navy had been working on a program called Ship Specifications for six years and had spent more than $600,000 dollars on it (approximately equal to $4.5 million today). It was using different computer programs to handle discrete aspects of the overall design process, which in the navy were far more exacting than those used in commercial shipbuilding. Ship Specifications would ideally get all of those independent programs to work together in one interface. So far, attempts to integrate all of these programs had been unsuccessful. As a matter of fact, the navy had deemed the endeavor an impossible task. Raye was given the opportunity to make the impossible possible, but instead of six years, she would have six months in which to do it.

Raye began working on Ship Specifications in March 1971. She traveled by train to New York City to spend three days at the naval architecture firm M. Rosenblatt and Sons, the contractor that had been working on the program.

"Because I started at the bottom with computers, I knew how to tear them apart and put them back together," she said. "So there was an advantage in that, and I checked myself into a hotel in New York with the navy's money."

Where Raye felt frustrated and thwarted on the job in Washington, she found receptive cohorts at Rosenblatt who were eager to make the deadline to get this program working. Night and day, she worked on the Ship Specifications program, trying to debug it and get it working properly. In her spare time, she went to the theater and enjoyed the sights of the city. Back at Rosenblatt, she would call people she knew at Boeing Aircraft and other companies for advice on how to get the computer's high-speed printer to print with upper and lowercase letters (at the time, all caps was customary), and then in two columns, both per the navy's specifications. Raye figured it out, and said she dumped the program on a disk and brought it back to Washington.

"When I got back to Washington, I needed four hours of computer time a day to tear the system apart and reassemble it so other people could use the program," Raye said. "They cut the computer off at seven o'clock at night, and I would come back in at seven-thirty and turn the system back on so I could work until midnight. Then, I'd come back in the morning—and I never charged them an extra dime."

Eventually, Raye said, Wally told her she could not come in and work by herself at night. "I told him I had to come in at night to get the time I needed," she said. "And he told me I couldn't work alone. None of my coworkers would agree to come in without getting paid overtime. So I took David and my mother to work with me because I couldn't work alone."

It was 1971, and David was almost five years old. By then, Raye had taught him how to code in the beginner's programming language BASIC, so she would put him in a corner of the computer room and have him punch up all the government's cards.

"I remember going to the office a lot with her, but maybe not when she was trying to go debug this program," David said. "I remember going into the office and punching cards, and I know I did it more than once. She'd hand me a stack of them and teach me to run the machines. Apparently, the cards were really expensive, but she would say, 'Here, they're not offering me any help, so print away.'"

When David got tired, Raye covered him up with a blanket and he would drift off to sleep. She worked until midnight as her mother did crossword puzzles. Then, the trio left so Raye could get enough sleep to work her regular shift the next day.

"After a while, Wally said, 'Raye, why are you bringing your mother and son in here? They don't work here,'" Raye recalled. "And I said, 'Well you told me I couldn't work alone. So now I'm not alone.' He said, 'Why are you so determined to do this?' I told him, 'Well, you gave me a deadline, and the only way I could meet that deadline was to come in and work at night.'"

The truth was, Raye didn't believe she had any other choice but to come in late with her family at her side to show her boss what was possible. Perhaps, deep down, she was trying to show herself what was possible, too. Although Raye may have felt that Wally was determined to see her fail, the next day he assigned a staff member to work with her from 4:00 PM until midnight. David and Flossie stayed at home, and Raye met her deadline.

Then, Raye took what she believed to be the next logical step in her professional journey: she asked for a ship to design.

Wally told her she couldn't have one.

"I said to him, 'What do you mean I can't have a ship to design?'" Raye said. "You gave me this deadline to meet, and I did it. He said, 'Well this was supposed to an impossible task. Nobody believed you'd actually do it. So nobody's going to use it.'"

Raye was angry, but she said she didn't let it show. She said she told Wally, "OK," then went back to her desk and didn't want to talk to anyone.

"For some reason, approximately two weeks or so later, President Nixon decided to give the navy two months, instead of the usual two years, to design a ship," Raye said. "The admirals came to me and said, 'Young lady, we understand you've got a system to design ships. The president is giving us two months instead of two years to design a ship. We can give you a month. Can you do it?'"

Raye Montague had made an impossible task possible, and so she was given the ship she had wanted to design since she was a little girl. Could her debugged program handle the task? She was determined to find out.

"Yes, I can do it," Raye said. "Columbus Day weekend is coming up. I'll do it over the weekend, or I can't do it."

After hearing Raye commit to designing a ship over a holiday weekend, Wally told her to take the entire month she was offered.

"No," she told him. "Because if there's bugs in the system, then there's no point in me wasting that time and money. If I can't do it over the weekend, then I can't do it and meet the deadline."

What could Wally say to this kind of determination? All he could do was turn it over to the admirals, who began telling Raye what they wanted to have on the ship.

"They wanted high-powered engines on it," she said. "I put two General Electric LM2500 engines on it, and those were based on GE's large aircraft engines. These things had a gas turbine, attached fuel and oil pumps, and other things that can start and monitor the way the engine is operating. Remember, I don't have an engineering degree, but I'm doing this, figuring out how to get these big engines on this boat. They kept telling me what else they needed and it was up to me to put it together."

Raye went to the lead counter, which was used to punch the computations for the program onto cards. She put the cards in the system, started it, and then worked all day as the machine began creating what she hoped would be a ship design draft.

She had been out late dancing the night before at a sorority reunion. Although she was sleep-deprived, it didn't matter. She would push through her fatigue and do what the admirals wanted—and prove that the ship specifications program on which she had worked so hard was a viable method of drafting ships quickly and cost effectively. Raye stayed at work until just past midnight, when she let the evening shift take over. On her way out the door, she told them to call her if anything happened.

She went home, took a shower, and the phone rang. It was someone from the evening shift, telling her that the computer had stopped.

"I rushed back to the office, and when I got there, everybody was standing with their hands against the wall just like they had been caught doing something wrong," Raye said. "'Where did it stop? What happened?' I asked them. 'It just stopped,' they told me. So then I asked them, 'Who hit the stop button? Did you touch it?'"

Raye walked over to the printer and gathered up the reams of paper. On the last page, it said THE END. There was her ship, designed to the admirals' specifications, using that impossible computer program that she made work. In eighteen hours and twenty-six minutes, she had done what the admirals had given her one month to do, and which countless ship designers used to accomplish in a manner of years. The following day she checked the printout for errors, found none, and then asked a colleague to make three copies of it and to have them bound. When the colleague returned with three copies of her ship draft, Raye put them on her desk and made a flag out of the one of the computer punch cards that read:

> Announcing the birth of the first ship designed by a computer. Proud mother, Raye J. Montague. Gestation period, 18 hours and 26 minutes.

Then, she left to enjoy the rest of the Columbus Day weekend at home.

When Raye came back to work on Tuesday, she took the elevator up to the seventh floor, as she did every day, and found everyone in her office standing there waiting for her.

"Raye, we've been here since 5:00 AM," Wally said. "We came to help you. Where did it blow up?"

"It didn't," she said.

"Well, where did you run into trouble?" Wally asked her.

"I didn't," she told him. "Have you been to my desk yet?"

Wally walked over to Raye's desk and saw her makeshift birth announcement, then called the admiral's office and asked to see him right away. Then, Wally grabbed one of the three binders from Raye's desk, and then her hand. He was taking her to see the admiral with him.

"We walked into the admiral's office, Wally slapped the binder on his desk, and said, 'Here's your ship and here's the young woman who designed it for you.'"

There she was, a thirty-six-year-old Black single mother from Arkansas. And to hear them tell it, she had just revolutionized the way the US Navy designed its ships and submarines. The ship she had drafted would become known as the Oliver Hazard Perry–class guided missile frigate—or the "little ship that could."

This would become the inexpensive solution the US Navy had sought to bolster its aging boats. The Perry was swift, thanks to the GE jet propulsion engine that Raye Montague never forgot about when she spoke about her ship design draft decades after the fact. It was also an easily maneuverable boat that could serve as a general-purpose escort ship for larger destroyers, and protect amphibious landing forces, supply and replenishment groups, and merchant convoys from aircraft and submarines. Because they were easy to operate and well-armed, they could also become part of various attack groups around the world. After drafting the ship design, the next step was taking the binder full of computer printouts and turning it into a real ship.

The accomplishment was a game changer for Raye Montague, but the next steps would be up to someone else to finish.

The navy honored Raye with its Meritorious Civilian Service Award on November 2, 1972. K. E. Wilson, acting commander of the Naval Ship Systems Command wrote:

> When you were tasked to direct the resolution and to effect the implementation of an automated Ship Specifications System which has been under development via contractors and in-house personnel, you accepted the challenge with enthusiasm. The development of ship specifications is the singularly most important item in the design, construction, and repair of naval ships. You willingly worked day and night, over weekends and holidays, without compensation in order to review the design, investigate the conceptual basis, and test the system's mathematical logic routines and interfaces, and thereafter modified and reconfigured the massive ship specifications automation system. It is significant to mention that you competently redesigned the major segmentations, modularized them into efficient executable patterns, and upgraded the entire mega-system to operate on the new, third-generation, giant computer, the IBM 360/91, in addition to the small, originally intended, second-generation IBM 1130 system, on which the mega-system was too sophisticated to run cost-effectively. In executing this assignment, you directed the performance of hundreds of design engineers in the coordination of the input and feedback sub-systems as the total system began to function. Due to your astute judgment and technical expertise, the heretofore impossible task of producing in one volume the thousands upon thousands of items that comprise

a total ship's specifications was accomplished. This marked the first time in history that accurate standardized specifications could be found at one time, in one location. This achievement heralds a new era in ship design which will produce a total ship specification in eight weeks rather than nine months.

By your conscientious performance, you significantly enhanced the Navy's ability to procure better ships, faster and more economically. "Well done!"

While the commendation was a personal thrill, Raye's success was perceived as a threat to some of her coworkers. She said her life was threatened, and people—Black and White—harried her when it was announced that she would receive this award. David was young when this happened, but he said there was some hushed talk around the house about people at work threatening her to her face. Although it is unknown who was threatening Raye, Fuller said it was likely the people who worked on the project with her who, he said, were probably jealous that Wally was giving her all the credit.

Raye accepting Meritorious Civilian Service award with David in 1972.

"The fact that she got the award was valid because of what Wally wrote," Fuller said. "But if you had made a contribution to the project, you might have a few negative feelings if you weren't honored, too."

Raye talked to management about the tensions, and they took the threat seriously enough to move her desk and parking spot closer to security. Flossie still worried about her daughter's safety.

"People felt like I should not have received that type of award before a White person did," Raye said. "But I earned this award many times over, and I was going to accept it. The more awards or accolades you get, the lonelier it gets."

Five men also received the Meritorious Civilian Service Award that year. Because of the threats to Raye's life, she said the other recipients had a ceremony on one day, and hers took place the day after that.

"The whole auditorium was officers, who were there to keep these people who threatened me from stopping this service," Raye said. "So I had to stand my ground and be very gracious. I was not going to let them know that they got me down."

For Raye, it was a lonely feeling to have succeeded and still have so many struggles, and so much strife. "I felt like I carried the weight of women and minorities at times," she said. "Regardless of what happened, I always had to try harder."

A few days after Raye received the Meritorious Civilian Service Award, Wally called her into his office and asked her to sit down. "He asked me what I'd like to be when I grew up," Raye said. "By this time, I had been working for the navy for fifteen years. I asked him, 'You want the truth?' He said, 'Yes.' I said, 'I want your job.'"

Raye said Wally told her that there were a lot of people who wanted his job but she was the first person with enough nerve to tell him to his face.

"That day, he became my mentor and helped me plan out the rest of my career," Raye said. "We'd meet in his office every day at lunch and among other things he'd teach me the nitty gritty of ship design and construction and encourage me to lecture at the Naval Academy so I could raise my profile and change society's stereotypes of women and African Americans. We had to get me professional recognition and get me into professional societies and all this other stuff."

It was a turning point in Raye's career, and also in her relationship with Wally, and she was ready to see where her next steps led. Because of her work on the ship specifications system, in 1972 Wally nominated Raye for the Federal Woman's Award.

"Although she was not the original architect in developing the software or the data bank for the [ship specifications] system, she was instrumental in organizing and directing a plan to introduce the system to the cognizant NAVSEC engineers," Wally wrote in his nomination letter.

The plan Raye developed included developing manuals that were readily and easily used by engineers with little to no knowledge of computers, developing software aids to allow easy use of the computer, training NAVSEC personnel on how to use the system, and helping others use the system until it became second nature to them.

"In short, she rescued a system for which much money had been spent, for which there was no in-house implementation plan, and which would not have met the deadline date without her fortuitous voluntary assignment," he continued. "She reconfigured the system to perform within six months, thereby meeting the operational commitment for NAVSEC. It would simply not have been done without her combination of system engineering and programming abilities and her total commitment and perseverance in her task."

It's an effusive letter, especially considering that it was written by a boss who Raye said wanted her to fail. It's hard to know whether Raye embellished her story in later years to make it seem more remarkable that she succeeded against all odds, or whether Wally embellished his letter, perhaps out of guilt, so that he could be sure she got the credit she was due. Whatever the case may be—and it could be a little bit of both—Wally was making Raye Montague synonymous in NAVSEC, and the entire navy, with the words "outstanding," "instrumental," and "expert."

"She worked day and night, over weekends, and the Christmas and New Year's holidays without let up and without additional compensation to review the design, investigate the conceptual bases, and test the system's mathematical logic routines and interfaces, modifying and reconfiguring the massive ship specifications automation system with the finesse of a neurosurgeon," Wally wrote. "Such production of specifications will no longer be the basis of contractors' belated claims against the government for late or inaccurate information. It also heralds a new era in ship design which will produce a total ship specification in eight weeks rather than nine months. It marks the day when a complete ship specification will cost $67,000 not $200,000. It is a major step toward achieving the NAVSEC long-range goal of producing a ship design in six weeks."

Regardless of what Raye may have originally believed about Wally, he was someone who was fully aware and appreciative of not only what she had done, but what it had meant, and he was going to champion that.

"Mrs. Montague's efforts are readily recognizable as the critical element in the successful completion of the program which will help the Navy procure better ships, faster and more economically," he wrote. "This success is directly attributable to Mrs. Montague's perceptive analysis of the problem areas, indicative of her superb knowledge of the real power of digital computers, and the result of her aplomb and diplomacy with many senior officials and workers, contractors, and administrators to effect such resounding success. Mrs. Montague has since earned a promotion to grade GS-13 and received a letter of appreciation as a token reward for her work, which does not really compensate her for the marvel of systems engineering manifested during this period which will have an everlasting effect on the art and science of shipbuilding and on every naval ship that will ever be built."

After sixteen years with the navy, Raye Montague was going places. Due to her success with the ship specifications program, the Department of Defense established a Manufacturing Technology Advisory Group. Wally had given Raye books to read about the topic because, as he told her, in the next five years, manufacturing technology was going to be the way of the world. In Wally's mind, Raye would be the authority on the subject. The Department of Defense wanted Wally to be the group's representative from the navy, but he refused it and said it should be Raye. She became the lone woman in the group, with representatives from the army, air force, NASA, and other tech-heavy parts of the government.

"The group would go to all the services and look at their projects that required manufacturing technology to see if we could find ways to improve their operations," Raye said.

She recalled a trip to Wright-Patterson Air Force Base for a project designing titanium wings. During the visit, there was a NASA mission that involved sending a robot to the moon to collect soil samples. The robot's arm would not retract, and NASA engineer George Salley enlisted Raye to work with him to solve the problem.

"We spread out the data, and I can't say which one of us found the glitch," Raye said. "It was probably Salley, because it was his program. But seventy-two hours later, the arm retracted. When an article came out about it, I cut it out and sent it to my old teacher Mrs. Holiday, who was still teaching. I wrote, "Baby, you told me to aim for the stars and at the very worst, I'd land on the moon. Just so you know, I've been to the moon."

And she was continuing to soar. Wally encouraged her to lecture at the Naval Academy because the midshipmen needed to know about CAD-CAM, which she had helped develop. CAD-CAM was the next generation of the ship specifications technology that Raye had tweaked for CASDAC. The data in that program could be used to manufacture other types of parts for other agencies at a time when the government desperately needed to streamline its costs and procedures.

"All this dirty work, all this other stuff, the nitty-gritty stuff that I did, I never took the approach that this was not my job," Raye said. "I took the approach that whatever you're doing, I want to know about it."

9

Equal Opportunities

Now that Raye worked ten minutes from home, she could drive there for lunch to check on Flossie and David before going back to the office to work for a few more hours. She was protective of her son, and it was hard for her—at least at first—to allow nonfamily members to help her with him. But she could trust her mother while she was away, and now she was close enough so that she never had to choose between her job and being present in David's life. She was definitely an involved parent, he later said.

"As much as my mother was working, she'd get up, barely able to move, and she'd take me to bowling," David recalled. "Then she started bringing my grandmother. They would just make it a family thing because I had to be there early in the morning for Saturday league, and then every other Sunday I had a different league that met at eleven o' clock. She would go there and order pizza and we would eat and it was a family outing. People were like, 'Your family comes here and eats?' And then they were also like, 'Where's your dad?' You had to live with those questions."

David said that he used to get upset when people insinuated that it was absolutely necessary for him to have a father living under the same roof. He didn't see it that way. His mother was there for him. His

David's third grade class photo.

grandmother was, too. And they were both raising him to be responsible, polite, and well educated.

When Raye was at work, David said his grandmother would call him into her room and talk to him about the importance of faith. They would read psalms together or play card games. "She talked to me about the importance of making good decisions and working hard," David said. "She never said negative things about people. She told me to be careful, and she said she would be disappointed in me if I didn't try to do the right thing. That was a big deal. So up until the end, she always had a role to play."

In the mornings, Flossie would bring him to school. When she wasn't minding David, Flossie would spend time with a friend of hers David only remembers as Mr. Nelson. "He was a very nice guy," David said. "They were together for a long time when I was younger. They would go to concerts and movies and stuff like that. He died when I was in my early twenties, and it was around then that it finally dawned on me that this man was her boyfriend. I never thought of them as being in any sort of romantic relationship."

When you're young, life is full of mysteries like that and confusing things that adults say and do. For example, around the time that David was finishing his Montessori education, Raye told him that when he graduated and went to college, she would buy him all sorts of clothes and other items. "So, when he turned six, they had a little graduation," Raye said. "When my friend found out about it, she said, 'Well David, how does it feel to graduate and go to first grade?' And he said, 'I'm not going to first grade. My mama said when I graduate, I'm going to college.' I forgot to tell him that he had twelve more years of school ahead of him."

After Montessori, Raye sent David to Our Lady of Sorrows. It was right around the time that Maryland schools began to integrate. "They brought all these White kids out of the public schools into Catholic school so they wouldn't have to go to school with Black children," Raye said. "So all of a sudden, David doesn't want to go to school anymore. He loved school and I asked him what was wrong. It took him a while to tell me."

David was one of two Black children in the grade and was getting teased by his White classmates. Raye went to the school and talked to his teacher, a nun who said she didn't know why David was upset.

"David said, 'Yes you do, sister. You showed that film on Little Black Sambo,'" Raye said, referring to an old children's story that portrayed unflattering stereotypes of Black people. "And I said, 'You showed Little Black Sambo? Do you know that's been eliminated from the schools? You're too old to teach.' So I went to the monsignor and I told him that I wanted that teacher out of there, because the kids were calling my son Little Black Sambo after seeing that movie."

The monsignor didn't accede to her demands, so Raye moved David into an Episcopal school, where he remained through sixth grade.

As much as he loved and respected his mother, David said his grandmother was the bedrock of the household. "Sometimes maybe I told her more than I told my mother," he said. "My mother was big on accountability. I'd remember playing around looking for stuff. I remember one time I was just rummaging around in the house, messing around in closets. I had a coin collection at the time, and what I recall was that I kept it in my mom's closet because she didn't want them out or have my friends messing with them. I found some sort of document that day that had the name Weldon Means on it. So I asked Mom who he was."

David said his mother paused, shook her head, and told him, "It's my ex-husband." She was not forthcoming with anything else, he said, other than it didn't work out.

Most of the time David's discoveries were not that serious. He came across a staple remover, thought it was a bracket to hold dentures, and put it in his mouth. Or, he stumbled upon a fancy staple gun his mother brought home from the office, and then took it in the garage to fire staples everywhere.

"The day I did that, I remember my mother got home and she was really tired, so she took off her shoes, got out of her car, and wondered what was going on," he said. "I remember her yelling, 'Get down here and clean up all these goddamn staples!' So I got in trouble for that."

Another time, David locked a desk door Raye needed to access and lost the key. "I had to go into my savings to pay for a locksmith," he said. "I was about eight years old, and it cost about twenty-one dollars. Then I found the key the next day, and asked her if we could get our money back. She told me no. She believed in accountability. I appreciate that now. I really do."

What David lacked in male role models under his roof, he got from other families in the neighborhood. "One of my friends had a dad in the fire department and he taught me all about cars and mowing the lawn," David said. "Another neighbor across the street would talk to me about school and stuff. He had two little girls and that's how I learned about how parents treat little girls versus little boys—you shouldn't treat them different at all. There was another neighbor who would take me with him when he went golfing. So my mom was happy that I had a supportive structure like this around me. I don't think she planned it that way."

What Raye wanted was a safe neighborhood away from a city where racial tensions had been steadily increasing since rioters had burned and looted downtown neighborhoods after Martin Luther King Jr. was murdered. Nothing had been repaired since then, the murder rate increased, and the anger about a break-in at the Democratic National Headquarters was palpable.

"I had some friends who came in from St. Louis who had friends who worked for the Democratic Party, and they wanted to go down and see them," Raye said. "We went down to the Watergate on the day of the break-in. We visited with these friends and went home, and that night is when it happened. You talk about being right there, that day, and it happened that evening. Truth is really stranger than fiction."

The crime led to the discovery of multiple abuses of power by the Nixon administration and an eventual impeachment inquiry. It was a strange time in the city, and a strange time for a twenty-seven-year-old naval officer named Peter Bono to be finding his way in it. He joined CASDAC to work on structural detailing programs, as "sort of a freebie."

"The navy didn't have many PhD computer scientists, and I dropped in their lap," he said. "The way the navy works is, when you are about to go on active duty, you have a detailer that assigns you to a slot in an organization. Wally had a need for a PhD computer scientist, so they asked him if he wanted me. Somebody must have figured out that they could get me for less." Bono said he stayed at CASDAC for three years and "it was great."

"I was given responsibilities that most young people didn't get," he said. "I went to naval facilities, and non-naval ones as well, and looked at what they were trying to do. I worked on future science. I managed specific projects. I kind of jumped into the middle of CASDAC and never asked how all this ongoing activity got started. Certainly, Wally was a visionary, but there were others there too who saw what computers could offer in terms of getting rid of the drudgery of all these calculations. They saw how computers could allow you to model something and do multiple trials and see how they work. You could try out different things. The only way you could do it before was to build physical models to scale and take them to the Model Basin and run different waves at them to see how they performed."

Everyone at CASDAC had specific projects for which they were responsible, each with its own budget and resources. In an open-concept office comprised of cubicles, it was pretty easy to figure out who was working on what. The walls were only about four or five feet high, so as people worked, they could hear other people's conversations and questions about their specific interests related to using computer graphics technology in the design and manufacturing process. At this point, there were some multimillion-dollar contracts from companies such as Lockheed, but things were still in their infancy, and the group was trying to advance their concepts with expert help from outside.

"At this point, the ideas were way ahead of the technology," Bono said. "It was the 1990s before computers were powerful enough to do what we wanted."

Some of the design concepts CASDAC was trying to computerize included fairing, which is the process of designing a ship's hull as a smooth curve without bumps and ridges. Hulls need to have a smooth curve so they can travel faster and create less noise and turbulence to avoid detection from submarines.

"It used to be that architects were good at drawing those fairing lines, but one of the advances of the sixties was that an MIT professor named Steven Coons developed a formulation for creating a hull's curves, and Pierre Bézier created one for smooth lines," Bono said. "What people who got involved in CAD ship design did was translate those requirements for smooth lines into equations that could be solved."

Depending on the design problem each person was trying to solve, they had to know how to go out and get the right expert or experts to help them apply their knowledge to a big-picture solution. All of the work had to be done within a budget that was not that large, maybe a few million dollars, Bono said.

Raye Montague was used to this reality, Bono recalled, and over time he befriended her. Bono said that he didn't notice any overt racism toward her in the office, but perhaps he was oblivious to it, he said. "When I walked into the office, things seemed normal to me," he said. "She was the only female technical person. I accepted that as normal and didn't think about it."

Bono said he was about five to ten years younger than most of the people working at CASDAC. While his colleagues accepted him, he said he felt like he had more in common with the young scientists working over at David Taylor Model Basin.

"At the same time, I was being friends with Raye and spending time with her as much as I could," Bono said. "We would have brown bag lunches together because I liked her company. Others were welcoming to me, but I always picked her to sit with. I can't remember a single conversation we had, but she always gave me this warm feeling. She was a lovely person with a great smile. Maybe I asked her what she did in Washington? Maybe she asked me what I did with my wife in Washington? I can't recall."

Bono said he only had a three-year window into Raye's life, and it came after one of her biggest triumphs at the navy, although he was unaware of her contribution to the Ship Specifications program, or to the development of the Oliver Hazard Perry frigate at the time.

"She had a long and varied career," he said. "And I know that at this point in her life, she was able to do things that gave her the chance to go from invisible to highly visible."

Raye's home in West Hyattsville was her refuge from the stresses of work. It was where she unwound after a long day, first in David's company, then in Flossie's, before retiring to her bedroom for the night. Most neighbors met Flossie and David before they met Raye, because of her work schedule and occasional travel.

Neighbor Sandra Howell recalled meeting David sometime in 1972. He and her son, Larry, used to play together. "We lived a street over from them," said Howell, who is now a retired schoolteacher. "At that time, kids met on the street and played with each other. So that's how Larry and David got to know each other. Then I said I needed to meet David's mother. So I went over to meet Raye, but she was at work. I met Mrs. McNeel instead, and she and I became friends first because Raye was at work a lot."

Howell met Raye about two or three months after the boys started playing together outside. Raye was cordial, welcomed Howell in, and they talked about their lives for a bit. "We both had a policy that the boys could play outside—they were boys—but not at each other's houses," Howell said. "After Raye and I met, we decided it was OK for them to visit at each other's houses. Flossie always liked to know where David was and that he was OK, and I was the same way about Larry."

Howell said Larry spent more time over at Raye's house than she did. "He knew he was always welcome there, and he never came home and said 'Mrs. Montague wouldn't let me come over,'" she said. "Larry was never off limits in their home. And David was not off limits in my home either."

Like many other friends and neighbors who knew David, Howell said he was a good boy who was being raised well by his mother and grandmother.

"Raye was the kind of person who was interested in David being a great person, which he is," Howell said. "That's why Mrs. McNeel came up to take care of David. I had never met David's father, but I would visit with Mrs. McNeel and sometimes talk to Raye in the evenings when she wasn't working late, or going to a meeting for one of the organizations she belonged to. I was married at the time, so I tried not to infringe too much on her life."

A friendship between the two mothers developed, and Howell said she deeply appreciated their closeness, in part because of the way Raye treated Larry like he was her own son. Raye, for her part, approved of David hanging out at Larry's house because he had a father present.

"David didn't know how to ride a bike, and my father taught him how," Larry said. "That's how we met. He lived the next street over, and at first he was the kid who was older and couldn't ride a bike. He was an only child, and I had a sister, so we sort of related to each other in terms of not having other brothers. We found out we had some of the same interests, so we liked to do stuff together."

Despite a bond between them that exists to this day the boys, Sandra Howell said, were different as night and day, and that could lead to occasional tension.

"Larry was outgoing and aggressive," she said. "David was quiet and calm. Sometimes they'd fight, and Larry wouldn't let David cross the street. Larry was mischievous. He wrapped toilet paper around their tree one time. He said he didn't do it, but we know he did. David could be mischievous too, but not quite as mischievous as my son. Raye and I never let their quarrels get in the way of our friendship."

Where Raye was gregarious, Sandra Howell said Flossie was quiet and soft-spoken, often puttering around the house doing chores, cooking dinner, or anything else Raye needed while she was at work. Larry remembers hanging out at the house, which had a basement room near the garage where they could do "fun stuff" while Flossie made them french fries.

"Raye was glad to have her there," Sandra Howell said. "She was grateful for all the things she did for her, so much so that that bought her a fur coat and all these other special things. Having her there was a relief, and it helped her to do more with her job."

Larry said Raye and David lived in one of the nicer houses in the neighborhood. It was big and there were plaques and pictures of ships on the living room walls. He never understood the significance of them when he was younger. He also remembered that Raye had a computer in her house long before most people did.

"She was always very protective of David," Larry said. "And that was at a time when parents weren't quite as protective as they are now. I had a little more freedom to roam, and he didn't. Maybe it was because he was an only child. Maybe it was because of her personality. When we'd get in fights, she'd always want to try to find out what was going on, where other parents would let the kids work it out. But we were boys growing up, and the thing is sometimes there were times when we weren't talking because of something that had happened. We worked through it though." And when they did, it was like nothing ever happened. Larry would head over to Raye's house to spend time with David while the adults threw parties or played bridge.

"There was a guy who sounded like Martin Luther King when he spoke," Larry said. "She brought him around so David and I could have some culture. And if you closed your eyes, he sounded exactly like him. A lot of the people she mingled with were distinguished like that; they were on her level or above it. I think some of that rubbed off on David, because he grew up knowing education was important, and that's part of what drove him."

In time, Sandra Howell and her husband separated and she would confide in Raye about the experience and her interactions with other men.

"She was my psychologist," Howell said. "We would talk on the phone, but it was never about the subs she was building, or anything like that. It was always about the boys and our friendship. She knew she could always call me, and I knew I could always call her. Sometimes we'd talk until midnight."

—⊶⊷—

Raye worked hard, she was an involved parent, and she stayed up late counseling friends. She was an engaging individual who was engaged with all aspects of her life, and the navy selected her to work on its equal opportunity committee, which was something of a next frontier too. The navy understood that it had a problem with racism within its ranks and began looking into ways it could "develop a far greater sensitivity to all of our minority groups," according to an article in the branch's custom publication *All Hands Magazine*. Even after Raye's successes within the CASDAC group, she said she was treated as if she were a secretary, or as she put it, "the help."

"I remember one time I walked into a meeting and a guy told me he'd like a cup of coffee, as if I was there to fetch it for him," Raye said. "So I told him, 'I'd like a cup of coffee, too. Make sure mine has cream and sugar in it.' You can't have people say things like that to you and not be ready with a comeback. You gotta have something. Always."

The navy was also aware of discriminatory housing practices toward some of its men, and "the depth of feeling" among Black personnel about such realities. It assigned Lieutenant Commander William S. Norman, who was Black, to identify racism problems in the navy. Although he said that there was no place in the navy for insensitivity, he limited his search to officers.

What happened to other staff outside of navy offices was out of the scope of its control or interest at that time. "I would travel sometimes and it would be fifty guys and me, and the navy would put us up in hotels," Raye said. "I remember one time we were in Louisville, and because of my name, they must have assumed I was a White man, a Frenchman. So the hotel assigned two guys to a room, and when I realized what was going on, I said I couldn't share a room with these guys. So they told me there were no more rooms, but there was a Black hotel up the street. I told them to give me a pillow and a blanket and I'd sleep in the lobby instead. They wound up giving me the presidential suite."

The navy decided to get more progressive about its quest to provide equal opportunity. Norman knew that the navy's bad reputation in the Black community went back to the Spanish Civil War in 1898 and up to World War II, at least. He said that Black people, regardless of their education, training, and experience, were relegated to lower-ranking positions such as stewards, and that out of an officer corps of 80,000, there were only 509 Black officers. He said the navy needed to come up with plans, programs, and analyses of itself to see whether they were truly eliminating discrimination—and if they weren't, to find ways to fix bad practices that existed.

"Once we can guarantee equal opportunity, then we can provide educational programs to improve attitudes and race relations," he said, adding that the navy was a microcosm of the American community. Deep down, Norman said, he believed there would come a time when his services were no longer needed.

In *All Hands*, the magazine of the US Navy, one can see that the navy was trying to show that it was promoting Black officers and giving women opportunities they might not have otherwise had. By June 1972, the magazine was featuring the navy's first female admiral, and in the following month, they dedicated their issue to women in the navy. Most of the women featured were working as nurses, in recruiting, and in management and maintenance roles. By 1973, they were looking to not only open up more opportunities for women, but also give other people more chances at advancement.

"When you're very busy and an organization doesn't want something to work, they give it to a very busy person, hoping they will fail," Raye said. "So they put me on the EEO committee. They assigned me to it in addition to my technical work. So it got so that I was doing EEO stuff during the day, and technical stuff at night."

Raye sought out people working in mailrooms who had college degrees and looked for ways to help them become upwardly mobile in their chosen fields. She wanted people to be moving up the career ladder within five years, and she wrote a manual on how that could be accomplished. Then the navy asked her to focus her efforts on what she could do to get more women hired. She traveled the country to speak to women about pursuing their professional dreams within the federal government.

"A lot of women were encouraged to be at home," she said. "They did not feel like they should be able to earn the same money as men, but I told them I was a single parent and I needed the same money that the guys got, at least to a point. They were floored that I could speak up and ask for what I was worth. They were amazed that I had the audacity to say, 'No, I'm not going to do that.'"

Much of Raye's advice to these female audiences was grounded in Betty Holberton's early mentoring and her own experience. That's what made her presentations so effective. She was able tell other women how she fought her way up the GS pay scale and show them that there was a place for them in the government.

"I was traveling so much, I was changing cities so often, that sometimes I would wake up in the middle of the night and know I wasn't home, but I couldn't be sure where I was," Raye said. "I remember calling down to the desk and asking the clerk what city I was in, and she laughed at me. I guess she thought I was drunk. I finally realized that if I placed the telephone directory on the opposite bed, I could wake up in the middle of the night and see what city I was in. And that was comforting to me, because I didn't just feel like, 'Oh my God, where am I?'"

Her experiences in front of mostly female crowds weren't always perfect. "I went to Panama City, Florida, and Pascagoula, Mississippi, and I had David with me," Raye said. "They hired a White babysitter in Mississippi who had five children, and they would entertain David all day while I was working."

One day, she had David with her as she addressed a crowd. "I was speaking to the ladies and a man said to me that I shouldn't have my son with me," Raye said. "Keep in mind, I felt that by taking David with me, I was exposing him to different places and people. I was exposing him to the fact that people could have this bias toward working women. So I told the man that 'what you're saying is why I have him here. He needs to be aware of people like you.'"

Raye had been groomed to walk into a room, well prepared, and to surprise people with her knowledge, expertise, and poise. However, it could be surprising to some people, especially those with more conventional views of the role of women.

"I remember speaking in Montreal, and this man came up to me with his wife and asked me why I was doing this," she said. "He told me I had a pretty face, so why? I figured out that what he was saying was that I should be married. I told him that just because I had a pretty face, didn't mean I didn't have a brain, too. And his wife turned to him and said, 'Told you.' I'll never forget that."

While Raye was handling EEO affairs for the navy and speaking to women throughout the country about the job options they had within the federal government, the ship she drafted by computer was about to be commissioned into the US Fleet. Although she created the draft according to her admiral's specifications and was celebrated for it, she did not receive an invitation to see the ship commissioned at the Bath Iron Works in Maine on September 5, 1976.

"The rest of them would get the launching and everything else," she said. "Once I did the draft design, I went on to work on something else: the LHA-1, the Seawolf submarine. Then, they'd pass the design I just did on to someone else who would do the dog work. If you feel like you have to mother it, then you get stuck. You pass it on to someone else and that's how you get to grow with new things."

The actor John Wayne attended the Perry's commissioning ceremony as a guest of one of the dignitaries, and when the ship was briefly stuck on the slipway he emerged from the crowd, climbed the launch platform, and gave the frigate, which had just begun to move, a shove with his hand. To excited onlookers, it appeared as if one of America's biggest stars, the very personification of a cowboy, singlehandedly pushed a warship down the slipway. Although Raye was unable to see it, one imagines she would have been doubly proud to see her baby touched by a Hollywood hero.

"It's not really traditional for designers to be at launches," she said. "But everyone else got to crack the champagne, and I didn't get to see it. At the time it didn't matter, but now I feel like I should have been allowed to see some of my work. I feel that I missed a lot."

Eventually, she did get her wish to see one of her creations while on EEO detail. "I remember the captain was the commanding officer of the shipyard and they said they should entertain the VIP coming in from Washington with coffee," Raye said. "So I went in, and that captain was singing Raye Montague's praises, but he didn't think that the EEO person and the technical person could be the same. I told him that my trip there was as EEO, but that I was the same person."

Up until this point, Raye had not seen the LHA-1, which she had designed. "When Litton Industries had done the design work on this ship, they sent it to Washington and called it a perfect test case," she said. "So they assigned me to do an analysis of it, to see if this was really bona fide. So I went through it and said, 'Nope they can't unload this thing with bombs and stuff flying and dropping. They do it with all these different things that they have going on. Litton said they wanted another assessment, and they sent the data to people in Dahlgren, Virginia, who said they agreed with me."

Raye said the navy wouldn't pay Litton a dime until they implemented the changes she requested. Litton relented and the ship was built at the shipyard. The captain there asked her if she wanted to see it. "He called his aide and told them to bring me a hard hat and glasses and they took me to see the LHA-1," Raye said. "We got in an air-conditioned car, because it was August down there. And so there I am strutting aboard this ship in my high-heeled shoes, because I'm speaking. Oh they adored me. The aide asked me what I wanted to see first and I told them to take me to see elevator hole number ten because that thing gave me a fit. The ship had all the conveyor belts, elevators, and the ramps with the palettes. I saw it all, finally. And I could imagine what it would be like under fire."

That was the only ship she had worked on that she had ever seen or toured.

Back at the CASDAC office, Bono said that some people resented Raye, because she got credit for the EEO work she did. "They said that didn't take any special knowledge or skill to do that," Bono said. "But they failed to recognize that it takes a lot of skill to be successful in situations that require knowledge of people."

And if there was anyone who knew a thing or two about human nature, it was Raye Montague.

10

Love and Happiness

As Raye became the in-house authority on manufacturing technology and became more involved with ship design, she met a man named James Parrott, a manager in the weapons systems division of the Naval Ship Engineering Center.

"One of the guys, an engineer, knew him and said both of us were at the top of the line in the command, so he thought we'd be good with each other," Raye said. "I didn't date people with whom I worked, but this guy sang his praises and so I met him. I worked with the wife of the guy who was trying to set me up with James, and I told her to make sure this guy wasn't married."

Raye's coworker told her James was married, but in the process of getting a divorce from his wife Mabel.

Raye decided to give James a chance, and he took her to the theater for their first date. A week later, James called Raye and told her that he was going to California, and he wanted to see if she could drive him to the airport. She told him she would. James had a brand-new Lincoln Continental, and when he got to the airport, he gave Raye his credit card so she could fill up the car with gas.

"When it came time for him to come back from California, I went back to the airport to pick him up, and he told me the car was mine," Raye said. "I told him I couldn't accept the car. He had not held me.

He had not kissed me. Nothing. I didn't consider myself going with him. So I said, 'Just like that, you're going to give me a car?' And he said, 'Yeah, you like it?' I had two cars already. I had one car, and my mother had another. But he was insistent and I said OK."

Raye had been staying home with her son, David, who had chicken pox. While she was taking care of him, James asked to come by for a visit. It was near Valentine's Day, but Raye told him she didn't date during the week. Again, he was insistent, and he brought her a card that said, "If you decide you don't want to be my Valentine, would you at least do me one favor . . ." on the outside, and "Marry Me" on the inside.

"I told him I didn't like him and didn't want to be with him and didn't even know him," Raye said. "I told him he had to be crazy. But he said he loved me. Next thing I know, he's got David, who was six years old at the time, in his arms. David's grinning. I said I can't do that. My mother was entertaining her bridge group downstairs and James went downstairs to ask her for my hand in marriage. My mother told him, 'Raye doesn't love you, and you don't want to marry her, because she's not going to let you manage her money. And you're not going to change her. What you see is what you get. She can't cook. She can't keep house. She is strong-willed. You do not want to marry her.' He came back to me, and he said, 'You married everybody else and you've been hurt. Maybe you should let somebody love you for a change.'"

Raye stood firm and said no. But James was undaunted. He asked again, and Raye told him she didn't want a wedding or reception, but a ring and a piece of paper.

"What kind of ring?" he asked her.

Raye told him she loved diamonds and had lots of them, but she always wanted a two-carat solitaire.

"If you can't get that, forget it," she said. Raye said she believed she was raising a bar that James couldn't possibly clear. But he took Raye, Flossie, and David to a jewelry store and told her to pick a diamond, any diamond.

"I looked in the case and told him which stone I wanted. They took it back and reset it in a Tiffany setting, and then James put it on my hand," Raye said. "He said he was sorry because it wasn't exactly two carats. It was two and a quarter."

James not only wouldn't take no for an answer, he also wouldn't honor Raye's desire not to have a formal ceremony. He arranged for a wedding at the Cathedral of Santa Maria la Menor in Santo Domingo, Dominican Republic. Completed in 1540, the coral limestone edifice is the oldest cathedral in the Americas, and, at the time, was said to house some of Christopher Columbus's remains. It was a colonial showstopper, full of dark mahogany, silver, and jewels, and it was designed to convey that the Spanish were there to stay in the New World. But would James be there to stay in Raye's life? That was the question. He was doing everything he could to whisk her away and sweep her off her feet. Yet, here he was, doing his damnedest, on an island that held the belief that anything bearing the name Columbus was cursed. James would tie the knot in a church that held the explorer's bones, but he would reserve a luxury hotel suite that overlooked the sea, perhaps in hopes that he and Raye could look out over the horizon and see a beautiful new future, instead of doom. The morning after her wedding, Raye called her mother to let her know that she and James were husband and wife.

"She told me to enjoy it for as long as it lasts, because it wouldn't last," Raye said.

Six months later, Raye would have James committed to a mental institution. "I asked the psychiatrist what I did to him, because I had only been married to him for six months," Raye said. "He told me I didn't do anything; James had something coming on for years."

According to Raye, James had been diagnosed with schizophrenia. It can involve delusions, hallucinations, problematic thinking, and behaviors that harm daily functioning. James had been obsessed with Raye for years, and had pinned up pictures of her everywhere that he could. According to the psychiatrist, James believed that if he associated with Raye, then her successes would rub off on him.

"I had no idea," Raye said. "But the doctor told me again that it wasn't my fault. I recalled one day after we married, that one of the navy captains went up to James and said, 'So you're the one who married Raye. You better treat her right, or else you're going to have to answer to the navy.' I think that sent him off the deep end."

Raye plunged into work once again. She enrolled in ship design courses, though she never went to engineering school. Raye's successes roiled James when he came home after a month in the mental hospital. Then Raye discovered that James had been seeing his previous wife, who had given him a lot of money after their divorce. This windfall was how he flew Raye to the Caribbean for a posh wedding and honeymoon. When those funds dried up, James focused on how he could spend Raye's income. He tried to buy an expensive car in Raye's name, attempted to get Raye's name removed from their shared accounts, and endeavored to change the beneficiary of her life insurance policy from her son, David, to him.

Given the situation, Raye said she grew afraid for herself, David, and Flossie. James left, and David later said that he wasn't sure whether James left without putting up a fight, or whether it was a restraining order that kept him away. But once he was gone, Raye changed all the locks on the house—"It was my house, that I had bought before I met him"—and filed for divorce.

Both parties finally appeared in front of lawyers in Prince George's County, Maryland, about divorcing and dividing their property on August 26, 1980.

In Raye's deposition, she said that she and James married on May 11, 1973, the day after his divorce from his previous wife, Mabel, was final. Raye had met James three months earlier at work, and he began asking her to marry him two weeks after that. At the time, Raye said, James told her that he was separated from his wife and in the process of getting a divorce. Raye said she never knew for a fact whether James was separated from his wife, but she did agree to marry him a month after they first met. They began living together in March or April of 1973, in the house Raye jointly owned with her mother in Hyattsville, Maryland. Parrott's grown son lived with them for two months and celebrated his twenty-first birthday with them at the house. Raye baked a cake and threw a small party for him, she said, because he told her he had never had one.

"I was trying to be a good mother," she said.

James, meanwhile, said that Raye treated his son poorly and could not provide him with suitable accommodations. He said that she pressured him to adopt her son, David, who was then six years old. Raye disputed that, saying that it was James who wanted to adopt David so that they could all share the same last name. She said James began the paperwork but never finalized it.

Just as there was disagreement about what James's intentions with David were, there were also conflicting accounts about whether Raye gave James a gilded diamond ring that once belonged to David's father. Raye told attorneys it was a loan, and that the ring was meant for her son when he grew older. James said that Raye told him that after his death, the ring should be returned to her son. James had done work on it, he said, and even though he wanted to divorce Raye, he intended to keep the ring.

Both sides battled about gifts given and taken away, and thousands of dollars that Raye believed James took and that he believed belonged to him. That money belonged to her aunt and uncle, Raye said, a promise she made to them about investing in their liquor store in Little Rock. James took the $5,000 that Raye had promised her family, transferred it to an account in his name, then withdrew the cash and closed the account.

"That money belonged to me," James said.

"How could it belong to you if your wife deposited it in the account?" one of the attorneys asked.

James couldn't answer the question.

Raye pointed out that James had retired from the US Navy on psychiatric disability, and that he could be moody, difficult, and hostile without justification.

"First he wanted everyone to have full family meals together," she said. "Then he wanted no part of it. He would come downstairs at two or three in the morning, cook a meal, and then leave the kitchen dirty."

James acknowledged that he had an "equilibrium problem," but he attributed the collapse of their marriage to "problems of relating to one another that ended up in dissent and arguments." He said his ex-wife, Mabel, had taken care of his father and son while he was married to Raye. Then on August 11, 1974, James had told Raye he was going to spend a little bit of time on a cattle and pig farm he owned. According to Raye, he never returned, but she continued paying his automobile policy for the next few years, because their finances were so tangled at the time of his departure. In the meantime, James said that although it was not his intention to get back together with his ex-wife, Mabel, he was currently living under her roof, and had been for the past few years. When the divorce was finalized, James agreed to pay Raye the $5,000 he took from her. He also returned Dave Montague's ring.

"Mom always felt bad that she had three failed marriages," David said. "But it was not her fault that [James] had issues with her. He thought he needed to be the successful one. He was in good position, but he felt like he was second fiddle. The advice I've heard her give other women, especially women of color, is that a lot of men of color have issues with women being more successful. It takes a special person to be cool with a woman's success."

After Raye and James fought over the investment in her aunt and uncle's liquor store, Raye became a silent partner in the business. David recalls going down to Little Rock with his mother to visit family, and he would be put to work stocking candy and doing inventory in the store. "My great uncle ran that store until he couldn't," David said. "My mom's involvement in the store was how she was able as a single parent to send me to college and put some money away for herself. A lot of people were always curious about that, but she never told anybody what she was doing for years."

Raye was committed to her family, and David saw her loyalty to them growing up. It's a loyalty that he harbors to this day. Still, he said he looked around at other families in his neighborhood and saw people who seemed to have strong marriages. He said he wouldn't know what a real father-son relationship should be until later in life. Like Flossie, Raye was discreet about her romantic life, especially with David. That was easy to do when David was younger, but as he grew older, David began understanding that some of the men who visited his mother at home were her boyfriends.

"Sometimes it was difficult for me, because I didn't understand why she was going out with this person," he said. "I knew she liked going dancing and to fancy parties, so I know that part of this was that she wanted someone to go with her."

One of the men who stepped into Raye's life after her divorce from James was George Brown, a former air force police officer who got a mailroom job with the US Navy after he was discharged. Brown learned nautical drafting at night and was hired by the navy as a drafts-man a few years later. That's when good things started to happen for him, he said. The navy sent him to classes, he was promoted to elec-tronics technician and, after that, program manager. Then he met Raye Montague, who was a program manager, too.

"She was in another division, but because of my ability with draft-ing and other matters, she was always asking me questions," Brown said. "That's how we became close."

Brown was responsible for giving parameters for interior commu-nications and navigation on ships. Raye took those parameters and put them in the computer to create a ship design. If she wasn't sure about something with the design, she'd catch people in the hall and ask them questions, Brown said.

"She was a strong individual," Brown said. "I've never met anyone in my life that strong, and that would include my mother and father. If she made up her mind to do something, there wasn't anything or any-body that was going to stop her. That's my whole concept of Raye. She was inspirational, too. I don't care what your job was, she was always with you, improving you in some way or another."

Brown said "the racial issue was always there" in the office, and the few Black employees that worked there were very close. "As a group, we worked through it by helping each other and holding each other to a standard," he said. "You always had to be a little bit above average to get any recognition. As a group, we supported one another to make it through it. All of us had to compare notes about what was going on around us so we could be sure of what was really going on. It could be isolating, but Raye just led us. She was like the Pied Piper."

Brown had also been through a divorce by the time he and Raye met. Although their relationship began on friendly, professional terms, in time it developed into something more. Brown said at some point "it felt like I had been knowing her all my life." Maybe the sparks flew at a Christmas party, maybe it was at a meeting. Whatever it was, they began spending time together outside of work.

"In every way, she made my life a lot better," Brown said. "She was a beautiful woman, bright and fun. If we had gotten married, I proba-bly would have ended up being a doctor."

Brown said he and Raye had discussed marriage at one point, but because of their previous experiences, they didn't want to take that chance again. They got along well, had fun together, and maintained what would become a close friendship over time. Although David said it could be difficult for him to tell when his mother was dating someone, Brown became a kind and gentle presence in his life, occasionally taking him to school events as a father might.

After David finished sixth grade, Raye enrolled him in St. John's Military Academy, where he was accepted because of his high test scores. A neighbor boy drove David to school for a little while, but then stopped, "because I guess he thought it would bring his cool factor down to be seen with me," David said.

Or it could have been that the neighbor's German shepherd charged at David one day, knocked him down the front steps of the family's home, and then bit him on the face. "I tried to fight him off, but a neighbor had to help me," David said. "I had teeth marks on my head for a while after that. They had to have the dog tested for rabies, and then I never saw the dog again. I don't know if they had him put to sleep or what. After that, the neighbor didn't drive me to school anymore."

By ninth grade, David was unhappy at the military academy. His teachers told him and Raye that he was not college material. "I said, 'What do you mean he's not college material?'" she said. "He scored ninety-fifth percentile when I brought him here. What have you done to him in three years that he's not college material?'"

According to David, St John's had a history of telling people from single parent households, most of whom were Black, that they weren't qualified to go to traditional colleges. The school then told these students that they should go to a vocational school, a community college, or not go to college at all.

David said he did have a problem remembering when assignments were due, and he knew he wasn't the best student. He also knew he had terrible handwriting, so for one report assignment, he decided he would type it instead. When he turned it in, his teacher gave him and F and sent him to the principal's office.

"I didn't know why they sent me there, but then they told me I had been cheating," he said. "So they called my mother at work and she told them she'd be right there. They accused me of not being capable of typing my reports," he said. "But my mom had a Smith Corona typewriter at home, and I would get up super early to type my reports. She would see me doing it, because she was getting ready for work as I was trying to crank out the last part."

David said his mother listened to the principal as he sat there disbelieving "I remember her looking down for a second and then asking, 'Did y'all ask him if he typed his reports?' They told her no, but then they said I hadn't demonstrated that capability before. So she said, 'Yeah, well did you ask him?' They said no again. So she said 'Well, ask him.'"

Raye was at an impasse with the principal. "She said, 'Well what kind of typewriter does your secretary have?'" David said. "The secretary was right there and said she had a Smith Corona. I told her that was the same type of typewriter I used at home. They were surprised that we had one. Mom told them she brought it home from work, because it was just like the one she had the office. Then, she looked around, grabbed a piece of paper, and told me to go over to the typewriter and type."

"What do you think you're doing?" the principal asked Raye.

"You accused my son of cheating, so he's going to go over there and show you that he actually knows what he's talking about," Raye said.

David went to the typewriter and typed. The principal apologized. And then Raye told them she was taking David out of the school.

Raye took David to the Kingsbury Testing Center. On the way there, she told David that she knew he was smart, and that's why she was taking him to see some people who would test him because she didn't believe what the people at St John's said about him. She told him if he did have a problem, they would deal with it. But there was no way they were ignoring it.

"I told the people at the testing center that if I was pushing him too much to let me know," Raye said. "But if there's nothing wrong with him, I'm gonna keep my foot in his behind and put him in all the right schools. They had him in there for two hours writing essays, doing math problems, everything. He did fine. St John's tried to destroy that boy, because they knew he was going to be a leader."

David said he remembers scoring high in several areas that day, but he discovered that he had a problem paying attention because he was interested in a lot of different things. "Paying attention to a lot of different projects seemed like a normal thing to me, rather than focusing on one," David said. "They told my mom that I would probably have a hard time figuring out what I wanted to focus on and that it would be frustrating for a while. They were right. I don't remember what I did for the next few days, but my mom found small private high school called Cromwell Academy, and that was the best thing I think she had ever done for me. It was a much more nurturing environment where I could be myself."

As a teenager, being himself meant a growing interest in girls.

"David was young and silly at the time," Brown said. "He would get into little things with some of the neighborhood kids, little aggravating things that they'd wind up laughing about at some point. And I seem to recall that eventually there was this one young lady that was leading him in the wrong direction."

David said the girl in question went to the Duke Ellington Performing Arts School. She was very talented and very beautiful, so the duo spent a lot of time together. "One night I went with my mom and grandmother to support her at this fashion show she was in," David said. "I remember we were walking toward the building where the event was and [this girl's] mom was in front of us. So her mom and my mom start talking and somehow the conversation heads into them talking about how we were spending a lot of time together. I don't know what led up to what, but her mom said, 'Just make sure your son doesn't get my girl pregnant,' and then my mom said, 'Well you need to tell your daughter to make sure she isn't screwing without a rubber.'"

The two moms started arguing, and David said he couldn't bear to look at his grandmother because he knew she had to be mortified by the exchange. The girlfriend's mother spoke ill of David, and then said she had to be more careful about who she let in her house. She was always talking about the nice things she had, from her house to her Cadillac and other possessions, and David said he didn't understand how she had all of those things because she was a single mom and he never saw her working, especially the way his mother did.

"She was always at the house, eating, all the time," David said. "She would send us out to get her food. I don't know whether she had an inheritance, or what. And I didn't ask. But I knew something wasn't right. She was focused on stuff and people taking advantage of her. I remember my mother telling her, 'We have our own stuff. We don't need your stuff. If you're going to be this way about your stuff, then don't let David in your house.'"

Like a mother lioness, Raye protected her boy. No one was going to villainize him, no matter how hard they tried. Then again, as David began dating, she also wanted to be sure he didn't make the same mistakes in love that she did. After the inter-mama skirmish, David's relationship with this girl continued for a little while longer, but eventually things petered out, as most teenage love affairs do.

"When I was dating, especially as I got older, Mom talked to me about being sure someone was 'the one,'" David said. "She probably envisioned me trying, and trying, and trying again, just like she did."

On the other hand, she had to respect a young man who continued to try, whether it was romantically or academically. Raye herself had spent a lifetime of trying, always with her mother's support, no matter the outcome. It's little wonder that, in gratitude, she supported the people and places that nurtured David as he came into his own. She donated her time and money to the Cromwell Academy because it welcomed David and empowered him to become who he was meant to be. "She even wound up being the speaker at my graduation," David said. David excelled at Cromwell, and because his graduating class was so small, he was able to speak at his graduation too.

"[David's] father and grandmother from the other side showed up at his graduation," Raye said. "None of them had ever played a part in his life. David got up to speak and he said, 'I wanted to thank my mother and grandmother who made everything possible for me.' And that whole side of the family is setting there, and I never told him to say that."

At the time, Dave was living in a housing project, and Raye would put together care packages full of food and ask David to deliver them to him. He had fallen on hard times, and as empathetic as David tried to be about that, it was difficult given his father's lack of involvement in his life. David's half-sister Debra said she dropped Dave off at his graduation, and Raye wanted to know why she wouldn't just come in. Debra said she could not be in her father's presence. He was just too hard to get along with, and so she left. Raye had purchased David a brand-new Volkswagen Scirocco for graduation, and Dave wanted to go for a ride in it once he saw it.

"Dave was down on his luck, but he didn't support David," Raye said. "He was going from woman to woman, making good money but fanning it out to everyone. So he had nothing, and he wanted to ride in David's car. David took him to a bus line and the rest of us rode home in my car. But his father should have gotten him that car. He'd done nothing else for him. I paid cash for the car because it was what David wanted."

Thinking about Dave's absence in her son's life, Raye got angry that David took his father for a ride in the new car. "Normally I didn't let things bother me," Raye said. "But when David came in, I said, 'What'd your daddy give you for graduation?' He said, 'Nothing.' I said, 'You mean he didn't give you a car or anything?' And David said, 'Well, Mom, you know his health hasn't been too good.'"

Despite being forgiving in this instance, David habitually referred to Dave as his "sperm donor," which upset Raye. "He would say that, and I would say, 'Well I loved your father very much, and he loved me. . . . God sent you what you needed David, even if the devil had to bring it,'" Raye said. "'I could have married someone else and had a child, but it wouldn't be you. I thank him for you. You are very special.'"

It is certain that David felt the same way about his mother. As for Dave, the sicker he got, the more people abandoned him, including his girlfriend at the time. But no matter their history, Raye did not turn her back on him.

III

BRINGING IT FULL CIRCLE

Raye was often invited to speak to local children
about doing well in school.

11

Another Direction

Raye learned at what she called "the school of Wally" for eleven years, soaking up Dietrich's knowledge about ship design technology, and then writing and speaking about it around the country.

"When I was making presentations for the Department of Defense, there was a guy who would always ask me if I had an engineering degree," Raye said. "He did this to me three times, and so I went to Wally and asked him whether I should get an engineering degree because this guy kept minimizing me. Wally told me, 'Oh no, you can teach engineering. You don't need to go back to school.' Wally said there were people who went to engineering school who couldn't pass the certification exams, but he bet I could do it."

Wally was right. Without that formal degree, Raye passed the certification exam and became an internationally registered professional engineer in the United States.

"I helped develop CAD/CAM, which was something engineers needed to know in order to pass the boards," Raye said. "They had to be proficient in manufacturing technology, but then these boards started asking 'Well what about the people who developed this stuff?' So I was grandfathered in as a registered professional engineer in the United States. One of the guys who worked in the industry said he heard I was a registered professional engineer in the United States, so why didn't I try for Canada too and see what they said."

Raye submitted paperwork for her second certification, which entailed getting several people to certify that she had been working in the discipline.

"So my license came back saying 'based on *his* qualifications, Raye J. Parrott is authorized to use RPE,'" Raye said. "My admiral came to me and said he was in charge of ships and subs and didn't have any registration, and that I had two registrations, one of them international, and I was the only one on the East Coast who did.'"

Raye eventually encountered her heckler again. This time, when he asked her about her engineering degree, she stared him down and told him she was certified internationally.

"And then I said, 'Next question.'"

Raye had developed the clout she needed to address the midshipmen at the US Naval Academy and the American Society of Naval Engineers. She spoke to them on April 11, 1979, about the promise of the CASDAC project, which was still all it was that time—a project. Raye stressed that the system was a viable tool, and, in fact, a way of life that ship designers and builders could turn to for cost-effective and solid ship design. The Naval Command recommended full implementation, she said, in what was to be an eight-year, $60 million program. The chief of naval operations had already approved the recommendation, Raye said, because of years of studies showing cost savings and efficiencies. When the project was finally given the go-ahead, it was slated to be a nine-year, $80 million program due to inflation.

"The survival of the shipbuilding community and our national productivity is dependent on design engineers and manufacturing or construction engineers working together early in the design process," she said. Doing so, and doing so with CASDAC-generated technology would be a benefit to the navy and to shipyards throughout the United States.

Raye was confident, in part because of Wally's support and mentoring. Learning at his side, she had become the authority on computer-aided design and construction, and a well-prepared one at that. On the rare occasions she didn't know the answer to a question, she knew how to collect herself before offering a well-considered response.

Raye accepting a Society of Manufacturing Engineers Award, after she earned her professional engineering license.

"It was just normal for me to sit up and speak," Raye said. "After being on the debate team in high school and college, I knew how to think fast and on my feet. I used to smoke, so when they'd ask me a question and I didn't have an answer, I'd light up a cigarette, take two drags, and by the time I was done, I'd have an answer. I bought myself time to think."

Wally had done everything he could to prepare Raye to take over his job, just like she wanted. For her part, Raye tried to prove to him that she was well worth his time and trouble. Despite her accolades, she looked at her recent performance reviews and saw that as much as Wally said he thought of her, and as well as she performed, he had only been marking her as a satisfactory employee. Frustrated, and fresh from her address at the Naval Academy, Raye met with him on April 19, 1979, her most recent evaluation in hand. Wally had marked her work as outstanding in twenty-six areas, and given her a satisfactory in six other categories, for an overall rating of satisfactory. After nearly a decade of diligence, she asked Wally what she needed to do to get an outstanding rating. Although she had received high ranks from NAVSEC officers, Wally's marks were what mattered most to her, and she wanted to understand what she needed to do to excel in his eyes, at least for the purposes of this particular piece of paperwork. Based on her notes from this meeting, he told her she needed to do a better job of meeting schedules and organizing her work.

Staying organized and on task might have been difficult for Raye as she juggled her CASDAC tasks with work and volunteerism for other agencies, companies, and community organizations. It was difficult for her to say no to requests for her time and talent. Any time a person or organization outside of the navy thanked Raye for her time, or honored her for her achievements, they would ask her if there was anything they could do for her. She asked them to write a letter to her admiral or commanding officer telling them about how grateful they were for her work or service.

"When they'd send it, the commanding officer had to put a cover letter on it," Raye said. "He'd send a copy to me, and then another copy went into my personnel file. So my personnel file became huge, and that's how I built my portfolio, with evidence like this that I did a good job."

Then Wally died.

"I called his wife to see what I could do and she said that he left instructions for me to plan his funeral," Raye said. "He heard me talk about the funerals I had planned for my family, so he told his wife to have me take care of things. She said Wally also had patents in his car, and that he wanted me to turn them in because they had been done on navy time and money."

Raye made the arrangements for Wally's funeral and wrote his obituary.* She said the funeral directors wanted to give Wally a lavish casket because they knew he was fairly well-to-do, but she told them that it wouldn't be necessary.

"They must have thought I was the help because they weren't talking to me," Raye said. "So then they asked his wife about buying

---

* Throughout her life, Raye enjoyed writing obituaries for people. She believed it was a service to the family and gave them a tangible recollection of a loved one.

the lavish casket and she said no, just like I had already said. Then his wife explained to the funeral directors that Wally wanted me to make all the decisions. So I told them I wanted to see him in an eighteen-gauge, sealed casket and to get him a vault, because that's how I buried my family. He died on a Friday, so we couldn't have the funeral until the following Tuesday. We got the notice in the Washington papers that Monday. We held a Masonic funeral for him, and at the visitation, I stood at the door with his wife and mother-in-law and greeted people from the navy and introduced them to the family."

Raye said that she and her mother wanted to sit in the back once the funeral began, but Wally's wife wanted them up front beside her. She and her mother moved up front to say goodbye to Raye's mentor, champion, and friend. Wally's boss, Nat Kobitz, came up to Raye at the end of the funeral and told her to take a few days off because she had been so busy helping Wally's wife tend to his affairs.

"He told me I needed to unwind and process what had just happened because I had been a family member to Wally, really," Raye said. "So I said OK."

Nat left the service but accidentally left his coat behind, so he called Raye at home to be sure she could bring it to him when she returned to work.

Then, when Raye went back to work, she discovered that she would not be moving into Wally's job as she had hoped for and worked so hard to do.

"When I went back to work, lo and behold, the guys in the office had decided that they were gonna take charge," Raye said. "Wally taught me how to do the budget. I was doing all the requests for proposals for funding that were going up to the Hill and everything. They didn't know anything about that. But they decided they were going to take all the jobs I had and divide it amongst themselves and leave me out."

Raye said that once she realized what was going on, she went to Nat Kobitz's office, handed him his coat, and told him what was happening. "Nat looked at me," she said. "He didn't believe what I was telling him. He said, 'All right. You can't be serious. Things will work out.' So I was like, 'OK.'"

Raye said she went back to her office and noticed that the requests for proposals she normally wrote were being taken away for someone else to handle. She sat there, wondering what to do about it, pondering her fate now that Wally was gone. A request to speak at a local junior high school came across her desk, and she decided to take it, even though she had a mounting problem on her hands. Inspiring kids would take her mind off of matters, at least temporarily. Raye visited Kelly Miller Junior High School on April 17, 1980, and the principal wrote her commanding officer, Rear Admiral James W. Lisanby, in gratitude.

"Mrs. Parrott served as an excellent role model not only in her presentation, but also in her personality," principal Elaine D. Simons wrote. "The students observed immediately that she is a very intelligent, affable individual who is greatly concerned about their future. Mrs. Parrott is making a significant contribution in her visits to the public schools of the District of Columbia and we certainly hope this service will be continued."

Raye speaking to a class in Maryland about engineering in conjunction with the Links Inc. outreach program.

Lisanby wrote Raye to share the note and thank her for visiting the school: "Your continued outstanding contributions to the community and in particular to the youth are rewarding services both to yourself and the recipients," he wrote. "The enclosed letter from Kelly Miller Junior High School is forwarded with my appreciation for an excellent job. A copy of this letter with its enclosure will be placed in your personnel record."

No matter what Raye's colleagues were up to at work, her portfolio continued to grow despite them. She said the tumult in the office eventually got sorted out.

"When time came for funding requests, they came to me and asked me to do the work," she said. "I told them I wouldn't do it. And they asked me why not. And I told them that I had no idea what they were working on, so how could I justify it? Well they went marching over to Nat Kobitz and told him that I wouldn't do what they were asking me to do."

Kobitz asked Raye why she wouldn't do the budget requests. She reminded him of the conversation they had right after Wally's death. "When Wally died, they took everything and divided it up amongst themselves, and took everything away from me," she told him. "And they've had everybody go off and do briefings, but I know nothing about what they're doing. So therefore, I cannot justify their budgets, and I will not write up something when I don't know what I'm talking about."

Kobitz got mad and called a meeting. When the entire office assembled, he asked Raye whether there was any task that was taken from her that she wanted back.

"No," she told him. "They can have it."

"Well how are we going to get these requests done?" he asked her.

"I don't know," she replied. "That's not my problem. I don't owe you anything."

Kobitz told her he agreed. "He said if there was anything I wanted, I could take it back, otherwise the guys were going to figure out how to get those budget requests done," she said. "And that's the way they were left with it."

Raye would soon be transferred into another division.

With Wally gone, it was a time of uncertainty. And when Raye was feeling unsettled, she would turn to the psychic Jerome Groom to give her a sense of what might be on the horizon. Groom primarily worked as a custodian, but Raye and all of her friends visited him at his row house for answers about the future. David often sat in the living room sipping a Coca-Cola while Raye and Flossie met with Groom, who began his sessions by praying on the Bible, then meditating, before telling patrons what he saw. Raye visited Groom a few times a year, taking notes as he spoke. Between sessions, she would place checkmarks next to the predictions that came true.

Raye took five pages of notes from a visit titled "Spring 1981." On the first page she placed checks next to "jubilant about work and money" and "proud of plans." By this time, she was hopeful about a new job and her work life after Wally. Groom correctly saw that she would get a proposition that she would give serious thought, that she would get news of California, that she would be getting in a situation to organize something, and that work would be turning around in a good way. She was also told, again correctly, that she would be getting more money at work, that David needed extra lessons, and that she would get news of someone who works on a ship. She did not successfully meet a nice guy she could marry. It is unknown whether these meetings provided her with any comfort, but there are suggestions that as confident as she may have seemed on the outside, and as rooted as she was in the Catholic faith, there was still something within her that was unsettled and seeking.

Raye kept a folded-up newspaper clipping about a Duke University study about eight things one could do to achieve peace of mind, and highlighted passages that spoke to her. She shouldn't be suspicious and nurse grudges. She shouldn't live in the past because that would lead to depression. She shouldn't waste time and energy on things she couldn't change. She shouldn't withdraw during periods of emotional stress and shouldn't indulge in self-pity, but she should embrace the old-fashioned virtues of honor, compassion, and loyalty. She also shouldn't expect too much of herself, advice she didn't seem to follow. Finally, she should find something bigger than herself to believe in.

Raye loved to speak to kids and young adults about finding and achieving success. For the most part she spoke about finding success in science and technology fields, but she was able to draw from her own life to share universal lessons about resilience from which anyone could benefit. Perhaps knowing her story could comfort and inspire people was the recipe for getting out of her own head, her own worries. That may not have been evident to her at this time, however, as she was mostly focused on work and making sure that David was doing well at school and getting him well positioned to get into a good college. But she kept countless requests to speak at schools and in front of young professionals that she honored from this time, a sign that she was more than an engineer—she was an inspiration after whom young people could pattern their lives. Throughout her life, she spoke about how important it was to her to reach people that others had given up on and show them another path to success in a way that made them sit up straighter and made their eyes light up. Perhaps beyond the horoscopes she enjoyed reading and the psychic she enjoyed visiting, she found great purpose in helping others. For all the coworkers and friends from all walks of life and family, she admitted she was lonely.

"A lot of people I knew came from Arkansas to DC and then settled down with someone else," she said. "I had to go through a divorce, and all of a sudden I was left out. That hurts, especially when you need that support. You go from playing bridge together to not being invited to play."

Raye suspected one reason she was no longer welcomed among her bridge-playing peers was that some of the women thought she'd be after their husbands.

"Finally, I had to say to one of them, 'Honey, I wouldn't sprinkle salt on your husband because every one of them has already asked me out. I wouldn't have them. None of them,'" she said. "She didn't have to be concerned about me. But then when I got married again, they wanted me around. That hurts. And also, you find out the higher up you go, the lonelier it gets. You have very few friends, until you make your life beautiful in spite of it."

⸺⟡⸺

After Wally passed away, there was a CASDAC Operational Environment Study that looked at the technical problems associated with implementing CASDAC techniques in the navy, and in shipyards building navy ships. From the very beginning of CASDAC's inception, people wondered how this experimental work would be implemented at shipyards throughout the country. It was a struggling project with inadequate staffing and uncertain funding. There was no incentive for shipyards to embrace the technology due to shaky circumstances in the industry, and the uncertainty from the navy about the project in general. Technologies that CASDAC had developed had only been used on a limited basis, and Congress was reluctant to support more experimentation.

In August 1981, Raye was transferred from CASDAC to the Office of Maritime Affairs and Shipbuilding Technology where she worked as a computer systems analyst in the manufacturing and shipbuilding technology (MT/ST) branch. The assignment, which was originally supposed to last four months, had her working in familiar territory. She was originally responsible for coordinating the manufacturing technology for hull, mechanical, and electrical components on ships. The rest of the details were to be worked out later, per an official memorandum. One month before her assignment was to expire, Raye's boss, Raymond Ramsay, submitted paperwork that extended her stay in his department, despite the fact that there was no job available. On November 16, 1981, Ramsay wrote that he could justify the move with a classified job description for a general engineer.

Although the specifics about the job were not made clear to her, Raye accepted the new challenge. However, by the beginning of 1982, it became apparent to her that she had been misassigned and put into a situation that was going to derail the upward trajectory of her career. On June 17, 1982, her first performance review confirmed her fears. Ramsay marked Raye as "below target" on four out of seven professional objectives and alleged that her immediate supervisor gave her more lenient marks.

"Montague is not attuned to the levels of personal initiative required to support the MT/ST program office role," Ramsay wrote. "The inability to handle multiple tasks, or perform completed single tasks is related to aptitude and a lack of technical expertise. Follow-up actions are either weak or non-existent unless repeatedly 'prodded' by supervisors. In short, Montague is <u>mis-assigned and out of her depth.</u> With the background of a computer systems analyst, Montague has, and will have, difficulty in attaining credibility with others in her field (from industry and in-house). An outgoing personality is insufficient when the fast-developing MT/ST program must be built on credibility recognized by shipbuilders and manufacturers."

Ramsay alleged that Raye made excuses for what he characterized as her low performance. She told Ramsay that she often got dual directions from him and her immediate supervisor, Jerry Cuthbert. Those competing directions were confusing, according to her, and Ramsay said he made sure that Cuthbert was the only one giving her directions to "invalidate further excuses for low performance." Ramsay then recommended that she be reassigned to a more appropriate department.

In a separate memo, dated June 18, 1982, Ramsay said that he was bearing a workload he could not sustain for the coming year. "I am now forced to equate the reasonableness of my personal commitment to the Navy and the shipbuilding industry with the hazards to my health and the nightly impact on my family relationship [*sic*] which have been most meager, since I volunteered for this program last July," Ramsay wrote. "I am not interested in making my wife an early widow, and I am not interested in letting George Sawyer down in his mission. As you can see, I neither want to expire nor quit. . . . The neutral, sensible course is to seriously bring this issue to your attention and state that out of 110,000 people in NAVSEA, it would be unconscionable to retain the MT/ST program staff at this level with the present on-board aptitudes and capabilities."

Ramsay said that unless he got relief, he'd be forced to resign or request a transfer. He said it was not in his character to practice brinksmanship, especially "when I believe so strong in the Navy work I perform. . . . The Navy could never reimburse me for the time and effort expended in the past year, nor for the load which was shouldered from a non-responsive MT/ST program staff." At that time he asked to transfer Raye and another colleague of hers to a more appropriate position. That alone may have been reasonable under the circumstances, but the review, which remained in Raye's personnel file, was what stung most. Raye viewed the poor ratings as an attack on her integrity and livelihood, and she decided it was time to fight.

"May I say at this point, I have never received such an unfair evaluation in my entire career," Raye wrote in the right-hand margin of her performance review. "I have always been recognized for giving my all to the job and I have brought a great deal of good credit to the command. I object to the idea that I am not a good employee; the low rating is unjust. I request it be removed and that I be transferred ASAP."

Then Cuthbert, Raye's immediate supervisor, wrote a memo one week later on June 24, 1982, noting that Ramsay set aside Cuthbert's evaluations of Raye and her colleague Dr. Franz Frisch and depressed "their scores in the Merit Pay System to the lowest possible level for the express purpose of ousting these two employees." He urged the head of merit pay to overturn the scores, saying it was an "egregiously harsh manipulation of the Merit Pay System."

"Certainly neither Ms. Montague nor Dr. Frisch belongs [in this department], where the jobs to be done are unrelated to their skills and experience," Cuthbert continued. "To a great extent, the same is true of me. They were both thrust into their positions willy-nilly, without regard for their desires or expectations. Their performance has not been good; on the other hand, good performance in the face of persistent programming for failure would require truly heroic resolution and character, and I do not think that mere mortal unheroism is grounds for condemnation. Each of them is in a predicament, but their predicaments constitute NAVSEA and SEA 90 problems, not just personal problems. To force the burden of solving these problems entirely onto the backs of the employees would be an evasion of management responsibility—certainly unfeeling, arguably incompetent, and perhaps even dishonorable."

Cuthbert said that NAVSEA needed to take a "mature and managerially competent attack" on the problem. "Last autumn, Mr. Ramsay unilaterally compiled a list of 'milestones' without regard for their realism, considering only his preconceptions and rejecting information from others," Cuthbert wrote. "Ms. Montague's MPS objectives were dictated by Mr. Ramsay, with his references to his milestones. These objectives bear a superficial resemblance to the MT/ST program as it now stands, and I accepted Mr. Ramsay's assurance that evaluation of performance would be made through appropriate interpretation of the objectives in the light of change."

Cuthbert said that because Ramsay had expressed his desire to "use the system" against Raye and Franz Frisch and had demanded constant documentation of their always-changing task assignments, he now considered his trust of Ramsay naive. Raye and Frisch were accustomed to working on one thing at a time until it was complete or they needed a supervisor's input. Because Ramsay forbade them to look for outside feedback, and never was consistent about his expectations, he made it impossible for either employee to do their job successfully. Employees whom Ramsay perceived to have leadership's support got better evaluations than those in his crosshairs, Cuthbert wrote.

"This is not using the system," Cuthbert added. "It is exercising caprice. The MT/ST program is a house of cards. Not even Mr. Ramsay's extraordinary efforts can continue to keep it upright with any confidence. Even if the workers assigned to it were appropriate to the task, they are too ill supported. Morale is abysmally low." Cuthbert concluded that the purity of the Merit Pay System was in question as a result, and he asked that his ratings for Raye and Frisch be restored.

Enraged, Ramsay wrote in a letter dated June 25, 1982, that he would be happy to lower Cuthbert's MPS evaluations too, before requesting that he be transferred out of his department along with Raye and Frisch.

The problem would take some time to fix, in part because it would entail finding Raye a job that was suited to her talents. But the more immediate concern—that performance review—was something that Raye tried to resolve on her own. She asked Ramsay to reconsider his low rating and use the one from Cuthbert instead.

Raye said that if the matter couldn't be resolved between them quietly, she would consider including a long list of things Ramsay did to her that impeded, demeaned, and degraded her. Among those things: telling her she was unwanted in the office, questioning her technical expertise and ability to complete tasks, ordering her not to interact with anyone outside of the office who outranked him, denying her ongoing training, abolishing her function as the recognized expert on CAD/CAM within NAVSEA, and telling her she had no credibility. She gave Ramsay five days to respond in writing, otherwise she would file a grievance against him.

Rather than wait to see what Ramsay would do, Raye began writing her grievance on September 17, 1982, just to have it ready.

"At least since December 16, 1981, I have been misassigned," she wrote. "Mr. Ramsay's ratings are made in relation to merit pay system (MPS) objectives established unilaterally by Mr. Ramsay on December 18, 1981, after the originally requested detail would have expired. I contend that Mr. Ramsay lacked the authority to assign the objectives or to establish an MPS rating based upon those objectives and applicable to me."

Raye contended that Ramsay decided to establish an inferior performance rating for her long before she started working for him, perhaps because of what Cuthbert had written in previous memos. She enclosed a September 15, 1982 letter from Cuthbert which read:

> As early as last autumn, Mr. Ramsay told me of his intention to establish MPS objectives for Ms. Montague and Dr. Frisch and to adhere to these objectively rigidly, and, on finding less than satisfactory performance to assign low ratings. He rejected my arguments regarding the unrealism of some of Ms. Montague's objectives. His intention of dismissing the two employees was made clear to me. His words were, "using the system."

Raye wrote that Ramsay maliciously "established unrealistic objectives; disregarded my complaints to that effect, privately and publicly; and lied to me about his intentions regarding the objectives he established." Furthermore, he "endangered [her] professional reputation and future employment deliberately and maliciously." Raye asked that his MPS rating of her be set aside as unlawful or restored to the rating given to her by Jerry Cuthbert.

Raye's grievance was ultimately filed and accepted on October 20, 1982. Almost a month transpired before Raye called Fred Gros, who was responsible for addressing her complaint. She took notes in that November 15 phone call, in which Gros admitted something needed to be done, but he wasn't sure if anything was. He would look into it, he told her.

"Sometimes I would go home and cry like a baby," she said. "But nobody knew it. But I would tell my mother about whatever happened, and she'd remind me that I could do anything. So I'd walk in the next day and they'd say, 'How are you today, Raye?' and I'd say, 'Great!' You'd hear me saying that to everybody. I'd come back the next day and they'd ask me again how I was doing. Again, I'd say, 'Great.' They told me, 'Now you can't be great every day, can you?' And I said, 'You'll never know, will you?' That's what got me through."

12

The Mentor

Raye Montague had been fighting a lifelong battle against racial and gender discrimination that aligned with the civil rights struggle of the 1960s and the women's liberation movement of the 1970s. The fact that she was Black and female was, as activist Frances Beal wrote, a double jeopardy.

Beal wrote that Black men had been "emasculated, lynched, and brutalized" and have "suffered the cruelest assault on mankind that the world has ever known. . . . By reducing the Black man in America to such abject oppression, the Black woman had no protector and was used, and is still being used in some cases, as the scapegoat for the evils that this horrendous system has perpetuated on Black men," Beal wrote. She continued that while Black women welcomed the growing power of Black men, they also believed that Black women should not be overlooked or left powerless.

"If we are talking about building a strong nation, capable of throwing off the yoke of capitalist oppression, then we are talking about the total involvement of every man, woman, and child, each with a highly developed political consciousness," she wrote. "We need our whole army out there dealing with the enemy and not half an army."

In Raye's case, she had the role of sole head of the household, supported by her mother. She worked in an environment largely com-

Raye working with colleague Ernest Glauberson at the computer.

prised of White men, some of whom were trying to thwart her success. But, as Beal wrote, Raye lived in a highly industrialized society, where everyone, Black or White, needed to be well educated and technologically savvy. "To wage a revolution, we need competent teachers, doctors, nurses, electronics experts, chemists, biologists, physicists, political scientists, and so on and so forth," Beal wrote. "Black women sitting at home reading bedtime stories to their children are just not going to make it."

Raye, for her part, would be both, which seems indicative of a period when women were beginning to bring home the bacon and fry it up in a pan, to quote the old Enjoli perfume commercial that was popular in the 1980s.* Raye was known to sing along with that jingle,

---

* The Enjoli jingle is based on the 1963 Peggy Lee hit "I'm a Woman." The revised lyrics for the perfume commercial are: "I can bring home the bacon/Enjoli/Fry it up in a pan/Enjoli/And never, never, never let you forget you're a man/ 'Cause I'm a woman/Enjoli." As children of the 1970s and 1980s, David and Paige have never forgotten this tune. However, despite Raye's habit of belting it out, she always wore Chanel No. 5.

and she didn't see any reason why she shouldn't be an engineer and an involved mother. There was no reason for her to choose one way or the other. In some respects, the decision had been made for her the second she decided she couldn't remain married to Dave Montague. In others, the decision she made seemed to suit her well. It was her household, run by her mother, whom she trusted. She worked hard and lived life on her terms, without a man telling her how to spend her money or raise her son. The men in her life had not known better, as she likely saw it, so there was no reason to rely on another one when she could rely on herself and her mother just fine.

Raye didn't believe she should be relegated to the sidelines because of her gender, and she worked to ensure that other female colleagues of any race were treated with respect and afforded the same opportunities as men. She also felt a special obligation to mentor other women of color, because they faced an extra obstacle in the workplace and in life. "Fifty years ago, all women got suffrage," said Dorothy Height in 1970. "But it took lynching, bombing, the Civil Rights movement and then the Voting Rights Act of 1965 to get it for black women and black people." Height said that for all the advances Black women had made by the 1970s, they were still largely employed in fields, such as domestic services, that had a lower median income than others. In that respect, Raye was an exceptional figure, and most likely a difficult one for her White, male coworkers to tolerate. On the other hand, she was among the 70 percent of Black women with kids aged six to eighteen who worked. So, at a time when the women's movement was gaining steam, Black women largely felt alienated from it. It was hard for them to see the White, middle-class women in the movement as being oppressed when they started from a place of more privilege.

"The total involvement of each individual is necessary," Beal wrote. "A revolutionary has the responsibility not only of toppling those that are now in a position of power, but of creating new institutions that will eliminate all forms of oppression." Beal believed that Black women should take an active role in bringing about this new kind of society where "our children, our loved ones, and each citizen can grow up and live as decent human beings, free from the pressures of racism and capitalist exploitation."

Raye may not have followed the writings of Frances Beal, but she did take an active role in making sure her son and others in her midst were to treated like decent human beings and that they had a future full of possibility. She took being a role model to DC-area youth, professional women, and her son seriously. She spoke about her journey toward the nation's capital and overcoming the obstacles she faced there whenever she got the chance knowing that, as many advances as had been made for African Americans in the United States, there were still prejudices and structures that were making it hard for many Black people to succeed. It must have been exciting for young people to see a well-dressed, well-respected Black woman who had climbed the ranks within the navy and excelled. It must have been an ispiring sight for other single mothers who worked and worried about how to pay for childcare see that they were not alone; Raye Montague struggled to get out the door to be present for her son in the evenings or at special school events. There was both an element of "If I can do it, you can, too" and "Let me help you through this rough patch" to Raye's personal interactions with people. Her generosity, her smile, and her warmth were among the things that contributed to the magic of her reputation. She did work hard, but she was not all about work. Her humanity was what lifted other people up. Black women especially needed a champion like Raye. Not everyone would stick their necks out for them in the same way.

Trenita Russell kept hearing about this new person named Raye who was being transferred into her department from MT/ST. She wondered who this man was, and even more than that, why coworkers kept asking her to hand over his mail. All of it seemed odd to her, but she kept working and doing what she was told. After all, she was a secretary and single mother of two who had been struggling to make ends meet. Raised by her godparents from the age of two, Trenita was asked to leave home as soon as her godmother died. She turned to a nearby church for help, and one of the congregants there took her in. Trenita was doing what she could to survive in those days, she said, whether it was looking for families who would take her in or depending on any number of rides or cabs to get to her tedious, entry-level job in a navy file room in Crystal City, Virginia. She became pregnant with her first child shortly after she was promoted to receptionist. She didn't know that "burn" meant to make a copy; then again, she didn't even know how to make a copy. Even worse, she was afraid to ask for help because she thought it would make her look dumb. The typewriter ink made her queasy, she said, but she stayed composed, "because I had no choice."

"I remember working in the office and doing a lot of filing and people aren't very nice to you sometimes when you're at the bottom of the totem pole," Trenita said. She recalled being asked to file things in file cabinets close to the floor. As pregnant as she was, it was difficult for her to get up and down. "There was a lot of prejudice going on. They wouldn't have the White secretaries carry heavy things or walk around if their feet were swollen. But they didn't care. They had no sympathy for you."

Trenita said she wore slippers to work when her feet were too swollen to wear shoes. Although she had a gray fake fur coat to wear in the winter, she couldn't button it and was frightened that her unborn child would get cold.

"I didn't know any better," she said. "People made fun of me because I was so big. They said, 'You don't have a baby, you have a football player in there.'"

Trenita said she never would have gotten pregnant if her god-mother had still been alive. As soon as the baby was born, she got an $80-a-month apartment but neglected to tell her landlord that she had a child. A babysitter kept Trenita's son, Kevin, during the week so she could work, and she got him back on the weekend.

"I didn't have a car, and because of this arrangement, I didn't have to get up early every day to get him to the babysitter," she said. "So it took the burden off somewhat, but I'm glad I had my faith."

Trenita said she felt like she had failed God. She "messed up" and had a baby out of wedlock, and "could have done better about going to church." Then a few years later she had another child with the same father. He didn't ask her to marry him, but he put a beautiful opal ring on her finger. The relationship did not end well, and by the time Trenita's daughter was born in 1975 she said, "People were talking about me like, 'Does she really think we're going to give her another baby shower?'" Trenita needed someone to nurture her the way her godmother once had.

To Trenita's surprise, this Raye person she kept hearing about wound up being a woman who, though Trenita could not have known it at the time, would become a much-needed presence in her life. "She introduced herself to me, and she had such a friendly, welcoming smile," Trenita said. "She was very friendly. I mean, here's a lady, she's in a nice outfit, nails polished, high heels, hair done. She's someone you could look up to. Someone you wanted to know. I was intimidated by her at first, because she was obviously a hotshot woman. I can still see her snapping her fingers and saying, 'Look what you could get if you were lucky.'"

Trenita soon learned that Raye was not like most of the higher-ups she had experienced up to this point. She was down-to-earth and treated everyone with respect. Trenita eventually confided that a coworker had asked for all of Raye's mail to be diverted to her, and Raye appreciated her honesty. She went down to the mailroom and made sure all of her mail was brought directly to her.

"The thing is, Raye started off pretty rough, too," Trenita said. "She started off low in the government and she had a degree. She knew people could be hard on you when you were low on the totem pole, so she would talk to me and give me hints on how to do things, say things. So many things she would tell me. She was there watching your back, like I did her. And it paid off."

Trenita said that she had been timid for too long and let people walk all over her. Raye told her that had to stop, and she gave Trenita the encouragement she needed to stand up for herself when people tried to hold her back. More than a coworker, Raye became a friend.

"Raye knew nobody trained me or helped me," Trenita said. "She knew that they took advantage of me. People would call me up when I was on leave and ask me to do things for them. Raye was totally different. She used me in the right way. She would ask me to get her a carton of cigarettes, or pick out a birthday card for someone. She would give me all this money and trust me. Trust is big. Don't step on people when they trust you."

Raye invited Trenita to her house on several occasions, and their children got to know each other as they grew up. David became a big brother to Trenita's children, and Raye became "Aunt Raye." Trenita said her kids were always afraid of doing or saying the wrong things around Raye because she was always so refined. They didn't want to let her down, and she was willing to do what it took to help them get into college.

"I went through a lot with my kids, but they ended up being good," Trenita said. "I used to cry to Raye about it, because she knew that you had to work extra hard to make up for the fact that their father wasn't around and involved. I told the father of my children that he couldn't hurt me, but when you hurt my children by not being part of their life, that's when I hurt. She understood that. Now my kids show me so much regard. All my struggling paid off. That's the beauty of it."

⁂

After moving to Trenita's office in the Naval Sea Systems Command, Raye was promoted to deputy program manager of ships. "She wasn't going to apply for that job, but we told her to," Trenita said. "She waited until the deadline, applied, and then got her package post-marked so they would have to accept it. She did the interview and got the job. People didn't like it, and that's when you'd see the jealousy come out. They did a lot of things to her. They hid papers, didn't tell her about meetings, didn't give her information. So when she got that job, she kept a low profile and didn't tell everyone what she was thinking because she didn't want it used against her." Trenita became Raye's secretary, which must have been appealing to her after the way she had been treated by previous bosses.

Where Raye had previously had tasks taken away from her, she now had a staff of 250 people and five field offices reporting to her. Her $2 million in resources was its own line item in the federal budget. "This is where my business degree came in handy," she said. "Most people in this position didn't have that background. And because I was a debater in high school, people would let me do the talking because I could communicate and persuade them . . . and argue both sides."

Every month, Raye went to the Pentagon to brief the Joint Chiefs of Staff about ship design, construction, and CAD/CAM. Her boss, Captain John F. Leahy, had no idea.

"He called me in to say, 'Raye, don't you know you're not supposed to go over there without going through all of these wickets?'" Raye said. "And I told him, nobody told me that. They just told me I had to brief them. So he told me if I ever went again to let him know. I told him I was going every month. And he said, 'How long you been doing this?' And I said 'Eighteen months.' So he told me, 'Oh just go on, Raye.'"

Working for decent people made the long hours bearable, and Trenita said Raye and Leahy were a great team. "When I was Leahy's secretary, he would allow me to come into work late so I could see my children off to school," Trenita said. "I told him I would never use my kids as an excuse to take off work, and he told me that he now was afraid they would get sick."

Trenita said she was working late one night and had to call her sister to go over to her house and feed her kids. Trenita wasn't able to leave work until the next morning, when people were filing in at the start of the next workday. "When I got home, my son was sitting in the stairway," Trenita said. "I couldn't climb the steps, I was so tired. Then Raye called and said we didn't have to go back to work."

Another time when Trenita was working late, she called her son at home and told him to get some liver out of the freezer for dinner. It had not been the first time Raye had overheard Trenita telling her son to defrost that meat. Raye gave Trenita $100 and told her to go to the grocery store and feed her son something else.

"I thought I had to pay her back, and I made very little money," Trenita said. "She just told me I could pay her back if I wanted to. She was always my shero. I once told her I wish she had come into my life sooner. She replied, 'I'm here now, so take advantage of it.' She had this way about her. She was always smiling and had that twinkle in her eye. She reminded me of Della Reese, she used to smile so much."

Trenita said Raye taught her how to forge her name because there might be times when she was away from the office and need something signed. "I told her, 'How do you know I won't sign myself a promotion?'" Trenita said. "But she trusted me. She was the person at the top pulling you up. That didn't mean she didn't get on me. I was her secretary. She did it in a way that didn't hurt your feelings."

It was not uncommon for Trenita to answer plenty of calls for Mr. Raye Montague, and she said she was always sure to point out that *she* was not available. Other times, Raye's friends would call and ask to speak to Trenita's mother. "I was like 'My mother? My mother's deceased. What do you mean?'" Trenita said. "They meant Raye."

One day Trenita was in her office when the White head of clerical came in and began rearranging her desk. Trenita said she told the woman that the rearranging would make things more complicated because she was right-handed. When Trenita spoke up for herself, the woman began screaming at her. What the woman didn't know was that some of Trenita's colleagues were around to hear her tirade.

"I mean, her shoulder-length hair was swinging, and all I could think about was grabbing that brown hair of hers and slamming her face into my glass-topped desk," Trenita said. "But I talked myself out of it because I had kids. I almost cried that day. I mean, there's a full-grown woman and she was talking to me like this. One of the secretaries, who was also White, asked me how I could just take it. I told her I had no choice. I wanted to leave because I felt humiliated, but I couldn't."

In the meantime, Raye told Captain Leahy about the confrontation, and he came out of his office to tell Trenita, "No one gives you any wrath but me."

"The next morning, I noticed his office door was closed," Trenita said. "I didn't know why, but then the woman who yelled at me the day before came out of his office and rolled her eyes at me because Captain Leahy told her to get another job."

A few days later, Trenita said there was a going-away party for the woman and she chose not to attend. "But I did go up to her and tell her congratulations," Trenita said. "Then I said 'God puts people in our path for certain reasons, and I hope our paths never cross again.'"

Meanwhile, Raye knew Leahy was about to retire and hoped that once he did, she would be promoted to his position. "They didn't want to give it to me officially," she said, adding that she eventually became acting program manager of ships. "They kept having other captains come in on an interim basis, but I was basically doing the work of a program manager, even though I never got that title."

13

David

As a teenager, David Montague met people from the Morehouse College alumni group in Washington, DC, which he said had a strong presence in the area—and a lot of clout with the school when it came to recommending promising young men for admission and scholarships. Each year, the group arranged a trip to Atlanta, called the Midnight Train to Georgia, to bring prospective students down to the campus for homecoming weekend.

"So on that train ride, you hear the alums talking about what the school did for them, and you're meeting other people your age who are scared about the future and not sure about what comes next for them," David said. "When you get down there, you're immersed in this whole big thing. And it was a big deal, unlike anything I had ever experienced."

David was being recruited by a number of colleges who wanted to offer him scholarships to attend because he had good grades and extracurricular involvement. But after spending the weekend in Atlanta with prominent alumni, he had Morehouse—and only Morehouse—on his mind.

Morehouse is part of the Atlanta University Center, the world's largest group of African American institutes of higher education. Founded during the Reconstruction era to educate newly freed Black men and women, at the time the cluster included Clark University, Atlanta University, Morehouse College, Morehouse School of Medicine, Morris Brown College, the Interdenominational Theological Center, and Spelman College. Located on Atlanta's west side, all of the campuses were so close to each other they often shared facilities. It's a university cluster with a star-studded group of alumni and a focus on excellence and character that David found appealing.

"My mom didn't know it, but Morehouse was the only college I applied to," David said. "And when I hadn't gotten any acceptance letters, my mother wondered why that was. I told her it was because I had only applied to Morehouse. She was livid. When I got in, she told me how lucky I was that I did."

When David showed up on campus in the fall of 1984, he settled in to his dorm room with a former Cromwell Academy classmate and began the weeklong Freshman Experience, which introduced new students to the school and its traditions.

"Everyone who goes through that gets a sense of pride and connection to the school that they feel for the rest of their life," David said. "I also got to meet people from all walks of life and see that Morehouse was this great leveling place. There were people with no exposure to things, people with some exposure, and people who grew up being told that they were better than sliced bread. I knew I grew up pretty well off, but then I met people who had to scrape together everything just to be there, and also people who had serious wealth in their families. There were some people who had two cars on campus—a BMW and a Mustang—and they would wonder out loud which car to drive each day."

Beyond the obvious socioeconomic differences that David noticed or had shoved in his face, there was the feeling that anyone on campus had the potential to change the world. More than that, there was a history of making such changes happen. Among the college's most prestigious alumni were Nobel laureate and civil rights leader Martin Luther King Jr., *Jet* executive editor Robert Johnson, 1969 World Series MVP Donn Clendenon, and pastor Calvin O. Butts III. There was great power in knowing you were a part of such a legacy, and the structure of the place was meant to reinforce that. During Freshman

Experience, David said upperclassmen brought the entire group of first-year students to nearby Spelman College, the Black, all-women's school nearby, to be registered to vote by Jesse Jackson Jr. and Coretta Scott King.

"I remember we were in Spelman's chapel, and there were people there from Morehouse and Spelman talking about civil rights," David said. "A few minutes later, Jesse Jackson got up and started talking about all the civil rights leaders who came through Atlanta and Spelman and Morehouse, and the struggle he went through to fight for civil rights. He got us all riled up. And then he said, 'Who you vote for is up to you, but it's up to you to vote, so don't give that power up.' He told us we needed to be in a position to exercise that power though, so he was gonna eliminate that barrier on the spot, at that moment."

The upperclassmen in attendance passed out voter registration cards and pens to everyone in the chapel, David said. By the end of that session, there were about three thousand newly registered voters, all of whom could say they were signed up to vote by two civil rights icons.

Once David was registered to vote, it was time for him to register for classes. He didn't really know what he wanted to study, but he knew he had to pick a major.

"I was listening to people who said they were going to major in computer science or business," he said. "Computer science wasn't of interest to me, but I didn't know what was, so I just said I was majoring in business. Then someone came up and was acting like a big shot, and they said they were majoring in *international* business. I don't remember what I said after that, but luckily I took core classes so I didn't really hurt myself as I figured it out."

Raye never pushed David toward a specific area of study, but she counseled him that whatever he did, he needed to be sure he did it well. "She told me if I was sweeping floors, to do it well, and to start my own business," David said. "Mom always told me she'd support me, no matter what I decided to do with my life. All she wanted was for me to go to college."

David didn't live up to his mother's expectations his first semester at school. Put simply, his grades were just not good. David said he wasn't partying, but he wasn't focused, so his freshman year was a wash. His main preoccupations were dating, playing arcade games, and simply trying to find his way.

"I think I was lost," David said. "I knew I was going to college, but I didn't have any idea about how I was supposed to figure things out." He'd get reminders every so often from his family, who encouraged him to study and behave himself. His great-aunt Gladys was often realistic about what he was doing with his time. "She'd tell me, 'Look, I know you have thoughts, but you leave those little girls alone.'" David said.

Every week, Morehouse and Spelman students met in their respective chapels to hear announcements about the coming week, famous guest speakers, and important information about current events and other matters that were relevant. At first, David didn't appreciate the mandatory attendance; you had to be there every week to graduate.

He also wasn't focused on his business studies, he said, and didn't like talking about them either.* "I kept trying to find it interesting, but it didn't speak to me," David said.

He lost himself in letters and care packages from home. As he read their notes, he felt like his mother and grandmother had a harder time with him being away than he did.

"I took your grandmother to her bridge club today over at Emma Holmes on Irving Street, N.W.," Raye wrote. "Seems as though I will be running her all around town every Saturday but that's the only way she can get around (smile). Everyone asked about you and was concerned about the loss of your wallet. I hope these items being replaced will solve your immediate problems. I hope you are doing well in school and please pursue the tutors . . . and study, study, study! . . . I love you son!!! Take good care of yourself and call me soon!"

Sometimes David would call them once a week. Sometimes he would call them a few times a week. Then there were periods where he didn't call them for a couple of weeks and they would phone him and wonder what was going on.

"Most of the time, when I called them, I didn't tell them what I was up to," David said. "I just wanted to be sure that they were OK.

---

* David said it did not help matters that the business school dean mandated that all third-year business students wore suits to class. He didn't see himself doing that.

Though I don't remember calling home to get advice about how to deal with things, I always felt like I had support." Raye had the usual mom worries. She wanted David to make sure he got his brakes checked. She wanted to be sure he had his car insurance and Triple A card in hand. She'd also send fun tales from home. "One of my staff from Puget Sound Naval Shipyard, Seattle is flying in and bringing fresh salmon," Raye wrote. "He asked me to start the coals at 17:45, so they will be ready at 18:00. He is going to be our chef." Following that, she describes an evening full of guests, wine, and potluck sides. "Seems like fun!" Raye said.

Flossie missed having David around and told him that every time she cooked a meal, she had to remind herself that she was not cooking for him, too. "I hope you are eating well," she wrote. "But I know that nothing tastes as good as grandmother's [cooking] (smile)." Flossie never shied away from talking about the food her grandson wasn't eating. She too wrote to David about how "Mom's peons came over for a cookout. The chef with the apron brought a fresh salmon. He prepared it here, put it on the grill, and smoked it. Others brought other dishes such as Watergate salad, onion dip, spaghetti salad, and whiskey cake." She wrote of broccoli salad and garlic bread, punch and wine, and kettle chips. She sent pictures, just so David could get the full effect.

After taunting David about all the good meals they were eating, Raye and Flossie would send him care packages full of food because the dining hall options were so terrible that students used to sneak into Clark University's cafeteria just to eat well. Sometimes Raye sent David hams with detailed directions on how to cook and store it, "as you eat from it from time to time." *Don't forget about this ham,* she seemed to say, in her ongoing efforts to remind David to eat, work hard, and stay focused.

There were always gentle reminders to David to acknowledge people when they sent him anything, to check in with a cousin who lived nearby, or to ask for help if and when he needed it. After an automobile accident, Raye reminded David not to wait too long to get his car repaired. The accident was not his fault, but the other driver was stonewalling him about insurance. Raye thought David was being too timid about following up, and she was afraid that the other party was about to take advantage of her son.

"If you wait too long, people will get the feeling they have gotten one over on you," Raye wrote in a letter dated September 15, 1985. "You are to be furnished with a car since it was not 'your accident.' You should not be inconvenienced at all. If they can't deal with those facts, I will contact our insurance and have them call the family's insurance and I am sure they would not want that."

There is nothing worse than having a mother threaten to step in when you're far away from home, trying to find your way as a young man in the world. But this was the way David grew up; Raye taught him what she needed to teach him, gave him everything he needed, and if something went wrong, she was there and fierce, ready to protect her cub if he needed her.

"She always pushed me to make sure people didn't get one over on me," David said. "She rarely offered to step in on my behalf, unless I was having difficulty navigating the process. She'd give me guidance on how to do it, but then she'd say I'd need to go do it."

Of course, Raye reminded David he should always be hard at work, treating school as if it were his job. Raye jokingly referred to school as "Montague Productions" in her letters; she always wanted to know if business was good.

Flossie filled the pages of her letters with laments about how much she missed him and the latest family gossip. "I talked to your Aunt Gladys," she wrote in a letter dated September 29, 1985. "Jones had gone to West Memphis to the dog races they race at night. He went on the bus. Gladys said he's so tall, he has to carry a small pillow to put on his knees to keep his legs from the seat in front of him. (smile). He won $300. He took off to the grocery store, bought ham, a side of bacon and several other things and Gladys asked him why he was buying so much. His reply was 'The dogs won't get this.' So he is really enjoying his retirement." Gladys, meanwhile, brought her dog back from obedience school early because she got too lonely without him, Flossie wrote.

Both women enclosed various amounts of spending money, stamps, and other tidbits in their letters to David. One wonders whether they coordinated the money they sent, or whether they loved David so much, they wanted to send him a little bit extra so they knew he would be happy and well cared for, knowing that his mom and grandmother were thinking about him.

"Take special care of yourself, Big Man," Flossie wrote. "I say a prayer every day for God to keep you in his care and I know that everything will be all right."

And, of course, there were the reminders: "Did you get your car fixed yet?" Flossie asked. "If not, don't wait too long. Their insurance might expire, so get it in."

He did.

As he struggled to decide on a major, David wandered the campus and observed the world around him. He found some kinship with members of the metro clubs that were set up at the Atlanta University Center schools. David said it helped him create more of a network with students from the DC area; they all organized to rent a bus that took them home for the holidays. That still didn't help him solve his problem about choosing an area of academic focus.

"One day I was walking down the hallway, and for some reason I slowed down and started reading a cork board with flyers on it," he said. "There were flyers about political action and supporting different causes and educating the community about things that weren't right." While David read the leaflets on what turned out to be a message board for the political science department, he wondered what the students would be like, and whether he'd enjoy the program.

"I remember hearing something and then realizing that it was some students in one of the professors' offices that were talking," he said. "I wasn't used to having interaction with my professors like that. So I stood around listening, and they started talking about local and international issues, and differences in philosophy around the world, and the ways that different types of people were being mistreated. There was a conversation about how to fight that."

David said he popped his head in the office and the professor invited him in. Some students had their books cracked open and were studying. Others were sitting on the floor talking to each other. As David took in the scene, he thought that this was what college was supposed to be like. He visited with this group of students for the next few days, never felt uncomfortable about speaking up, and enjoyed the range of their conversations. It felt like a safe place, David said. What's more, it was an interesting one. Hooked, he selected political science as his major. David became more engaged in his classes, and his grades went up as a result. He also started to appreciate the weekly chapel meetings he used to resent.

"I appreciated the logistics of getting these famous people to come talk to us, and the fact that they wanted to be there, telling us information that was important for us to know," David said. "If you were talking about how to make change happen, these meetings ingrained in you a sense of responsibility. And that's why I am committed to providing opportunities for others, because others did it for me."

His mother, grateful for the opportunities Morehouse was offering her son, became a contributor to the school and helped set up its own Navy ROTC chapter. Previously, Morehouse men who wanted to be in the Navy ROTC had to do so through the program attached to Georgia Institute of Technology. Raye thought they should have their own standalone chapter, and she worked with the navy for more than three years to make that a reality. The school finally got its own program in 1987, and when it did, the DC alumni chapter honored her at its annual regional conference for being an outstanding parent and friend of the college.

That same year, Raye rode the Midnight Train to Georgia down to visit her son during homecoming weekend of his junior year. At the time, David was living in old apartments that had been converted into dorm rooms. His room was a former kitchen, which had a back door that led out to a fire escape. The walls were still roughed out from where the appliances used to be, the fan was sealed up, and the floors were filthy linoleum. Before Raye arrived for the weekend, David went to Home Depot, bought the cheapest indoor-outdoor carpeting he could find, and secured it to the floor with duct tape.

"My mother came in and saw what I had done and thought it was the most ridiculous thing she had ever seen," he said. "But I said, at least I knew it was new and clean." Then Raye peered into David's study carrel and saw a sign that read KEEP YOUR BLACK ASS IN YOUR BOOKS.

"I was so embarrassed," he said. "But she told me, 'If this is what it takes to get you to focus, then great.'"

All signs were pointing to David's success. By the time he was a senior, he was an intern for Maynard Jackson, who was exploring another run for Atlanta mayor, and he was a delegate to the Democratic National Convention, which was held in Atlanta in 1988.

Raye was confident that she had equipped her son with the skills to survive and thrive in a city as large as Atlanta, but she did worry. All of the early lessons he had about taking care of himself in a society that was not inclined to treat him equally or fairly came into play while he was at Morehouse. David said that "strange things" happened to some of his classmates. There were murders and beatings that you'd hear about, he said. Day after day, he did what he could to stay out of trouble and not put himself in bad situations.

When David was working for Maynard Jackson, he said he was sent downtown one day to check the politician's post office box. He was struggling to find a place to park and asked a Black man who worked in one parking garage if he could leave his car with him for one minute as he ran inside. The man told David he would gladly watch his car, and David trusted him.

After running in to check the post office box, David came back out and asked the parking lot attendant for his keys. The attendant pointed him in the direction of a Plexiglass booth where a twenty-something White male was working. David said he went to the man, asked for his keys, and the man behind the counter told him that he would have to pay one dollar.

"I tried to explain that the other man told me he was going to watch my car as I ran in real quick, but the guy behind the counter was talking to me like I was an idiot," David said. "So I paid, and he just stood there. So I asked him for my key and he started messing with me. Eventually, he gets my key and slaps it down toward me."

David said the man then tried to instigate a fight. "He said, 'Oh you don't like that, do you? You don't like the way I just threw your keys down. You know how your people get. They get all mad,'" David said. "I was thinking to myself, 'Is this really happening?' and then the guy stepped outside of the booth and kept going. He was challenging me to a fight and I was getting mad, but something stopped me. I realized that if I crossed that line, he could tell people I was trying to rob him and it would be in the news."

David decided to leave. He asked the man for a receipt, and the attendant handed him an old one that had already been printed out. "Then he said to me 'I'm surprised you can read,'" David said. "I told him I was a college student. Of course I could read. So I went back to my car, pulled it around, got out of my car, and said to him, 'You may have gotten my dollar, but I got proof that you're a fucking asshole.' He turned beet red and then punched the Plexiglass and broke it. I told him, 'I betcha that will cost more than a dollar,' then got in my car and drove off."

David said his mother prepared him well for situations like that and surrounded him with people who helped him develop a strong sense of self. That, combined with five years at Morehouse, shaped him in such a way that he was eager to get out in the world and make a mark.

Graduation came and it was one of the hotter days on record, but the family came down to see David don a cap and gown and get his degree. Oprah Winfrey was the commencement speaker and she established the Oprah Winfrey Endowed Scholarship Fund with a $1 million gift.* Raye cried, in part because she was so proud of David, but also because the first class of Navy ROTC midshipmen had graduated from the school. Because of Raye's role setting up that ROTC chapter, she saw the young men as her sons too. That chapter not only enabled these students to go to school, but it added people to the military.

The event was also a little bit bittersweet. During David's sophomore year, some of his classmates were forced to leave school because President Ronald Reagan had cut funding for Pell Grants. David knew how fortunate he was to be a college graduate and to have come from a family that not only valued his education but could afford it.

---

* To date, Oprah Winfrey remains the largest individual donor to the school, with more than $25 million in total donations. She has put 600 students through school because of her scholarship.

After the ceremony, David said a lot of his classmates were going to chain restaurants to celebrate, but he didn't want to go somewhere and have to wait in line. He wanted to relax, and he chose to eat at the restaurant at the Holiday Inn.

"My mother was like, 'Are you sure this is what you want?' and I was like yeah," David recalled. David's entourage settled in at a big table and started to place their orders. One of Raye's friends started chatting up the server and asked him where he was going to school. He said he didn't have the money to go at the moment. Overcome by the optimism after the ceremony, the woman gave the server her contact information and offered to send him to college.

"I don't know if he ever followed up on it," David said. "But I remember how my mother's friend just sat there and we were all looking at her. She told us, 'I don't have any kids that are going to college and if I can do this to help someone, then why not?'"

Meanwhile, David planned to return to Washington to find a job. Raye wanted him to go right to graduate school, because she was raised to believe that the second you took a job, you'd get distracted and never finish your education. David told his mother that graduate school was in the cards, just not yet. He was going to find work, and he would eventually get a master's degree. For now, she would just have to trust him.

14

On the Shoulders
of Giants

Raye Montague—she had changed her last name back at this point—had learned a lifetime's worth of lessons in the navy and as a woman of color raising a son with her mother's help. Now that son was taking the lessons he had learned from her and was figuring out how to apply them in his quest to become a young professional who would eventually help people the way his mother and other mentors had helped him. The way Raye saw it, though, her work helping others was still not over, even with David making his way in the world. She once learned at the school of Wally and had something to show for it. Now she welcomed others into the school of Raye, hoping that they would achieve success from her teachings, too.

In an undated speech titled "Communications and Image" that she delivered at a government employees conference, Raye told her audience that success in the workplace came from demonstrating that you were a person of substance and integrity. Her remarks seemed ripped straight from her own experience, and were laced with self-deprecating humor.

"Most of you have heard that someone asked Cornelius Vanderbilt how much his yacht cost and that he replied, 'If you have to ask, you cannot afford it.'" she said. "Before you shed tears about a yacht you may never own, be apprised that boats, like children, require lots of care. Two of the happiest days in a boat owner's life often are the day he buys the boat and the day he sells it. The corporate corollary to the Vanderbilt yacht story is, 'If you have to ask how to become a vice president, then you cannot be one.'"

The road to climbing the workplace ladder begins with mentors, she said. However, she cautioned that you cannot pick the mentors you want. As she spoke, perhaps she looked back on Betty Holberton, who championed her early in her career, or Wally Dietrich, who at first seemed to single her out and want her out of his department. Each of these people saw something in Raye and empowered her to succeed. Recalling that, she found things in people like Trenita Russell and gave them the tools they needed to get ahead too.

"Mentors will pick you if you show promise," she explained. "Mentors are earned by hard work and can be lost by failure to listen, learn, and apply yourself intensely. In this case, you have to come to this conference to get advice on how to succeed and, since all the erudite people are out of town, it is up to me to give you some."

She said that coworkers would be making note of people's character, personality, and value, noting instances where one was proud, unprepared, and unwilling to put in appropriate effort. Everyone should be willing to improve on his or her shortcomings, she said, "otherwise the score will mount against you rapidly."

From the beginning, Raye knew what she didn't know, and she was willing to listen and learn from others so she could be given extra responsibilities in the office. She was also willing to put herself through night school to learn coding, and humble enough to know when she needed to ask someone else for help or fresh ideas. Perhaps some of this was rooted in this deep-seated fear that she would never be seen as anything but a Black woman from Little Rock, or perhaps it was rooted in her firm belief that a Black woman from Little Rock could do anything she set her mind to. Whatever the case, she was determined to show that anyone could excel if they focused on the task at hand.

"Rule two is from me," she said. "Solve your boss's problems, not your own. If you come to work with your problems on your mind, then your boss eventually will note that his interactions with you are not solving his or her problems, and he or she will avoid you. If you come to work every day after identifying and thinking about how to solve your boss's problems, your contributions will become recognized and often sought [after]."

How many problems had she solved for bosses over the years? When computer operators didn't show up for work one day, Raye did their job, mainly because she had been quietly watching and listening to what they did every day. Someone needed to twist those knobs and punch those keys, so Raye did it. Night shifts needed workers, so Raye drove a car she barely knew how to operate to the office so the work could get done. Wally needed someone who could make ship specifications software work properly and Raye not only worked out the bugs, but used that program to design a ship in less than a day. Raye knew from her own experience that higher-ups noticed—and sometimes rewarded—effort like she put in. It wasn't because of how she looked. It was because of what she did. Her actions spoke louder than words, color, or gender.

Yet, she always understood that no matter how hard she worked, there was always someone else out there who was just as talented and hardworking, if not more so. "To be promoted, you have to beat them," she said. "If you think that you can do that in a resume and an interview, you are in for a rude shock. . . . The subsequent results will indicate whether you simply lost a close race or have taken yourself out of the running. After two or three losses, you can accept your inadequacies or go back to rule one and start over."

Her ultimate point: "You can succeed if you expect to and work hard at it. Most people who attend college can learn enough to succeed if they try. . . . Remember, Einstein failed his college entrance examinations the first time, but he did not quit."

Nor did she. When she was seven, she didn't let the man in the submarine deter her from becoming an engineer. She didn't let Jim Crow keep her stuck in her so-called place in the Deep South, where she had few career options. She didn't let her college math grades, her more challenging bosses, her single motherhood, or anything else get in the way of where she wanted to go. When she spoke to crowds like this, she wanted them to understand that they shouldn't let their perceived obstacles get in their way either, no matter how hard the fight was.

Then Raye addressed a subject that was taboo for its time: mental health. "Your emotional strength peaks at approximately age twenty-six, as does your physical strength if you keep in good shape," she told her audience. "But while demands on your physical strength diminish after age twenty-six, the demands on your emotional strength usually increase. If you are not careful, the emotional demand line and your emotional strength line may cross. If that happens, the emotional overload will cause you to collapse."

Aside from the stress and strain in the workplace, there were the quieter challenges at home: The husbands. The son, frail at birth, who grew into an accomplished young man. The neighbors she had to win over because they thought single women were a recipe for trouble. Trying to be a good worker, boss, mother, daughter, friend, girlfriend, neighbor, while keeping a smile on her face as she did the work of showing up, day in and day out. She could go out dancing or play bridge and forget about some of her troubles, at least for a moment. But Raye knew that some things just lingered, and it was important to address those before they brought you to your knees.

Prior to the speech, Raye was given topics to discuss, among them dressing for success and presenting oneself well to a group. During the speech, she admitted that she wouldn't be talking about superficial things, because none of them would help you navigate a workplace full of complicated people. Talking about mental health in that particular moment in time seems revolutionary because of the stigma attached to it. Yet Raye Montague was not one to sugarcoat reality.

Raye was involved with the Links Incorporated, a nonprofit of more than sixteen thousand professional women of color. It is "one of the nation's oldest and largest volunteer organizations . . . committed to enriching, sustaining and ensuring the culture and economic survival of African Americans and other persons of African ancestry." David said it was common for Links members to talk about physical and mental health among its members.

Raye dressed up for a Links Inc. event, date unknown.

"They looked at the total aspect of wellness at the time," David said. "I think they did it from the standpoint of how some of these women would go out into the workplace and have so many people messing with them, that it would be necessary for them to think about themselves, and how to take care of themselves."

Raye's own struggles to survive and thrive at work perhaps offered some quiet glimpses of what this strong, self-assured executive grappled with when she was at home, away from coworkers' eyes and ears. "You can react to any psychological obstacle in five ways," she said. "Four are wrong. You may act regressively—act childishly. You may react aggressively, that is, attack others. You may be evasive—attempt some form of escape. Your reaction may be neurotic, that is, you exhibit behavior at odds with reality. That may or may not get you into a straitjacket, but certainly will interfere with your job performance. The correct choice . . . is coping with an obstacle coolly and rationally. In each instance, you must keep yourself in check and focus on finding a solution. Over the long term, you must also do something else."

Raye said that if you were going to "push your mental accelerator to the floor boards and leave it there all day, every day," then you had to find a way of leaving your job behind you at the end of every workday, so that you could be in a pleasant frame of mind when you got home.

David recalls that when his mother came home, she was always in a pleasant mood and always made time to ask how his day was before turning to Flossie to see if there was anything he left out or forgot. Flossie would then ask Raye about her day, and they would discuss it briefly, David said, the two of them talking through whatever may have happened at the office. On particularly stressful days, David said his mother would fill a large bowl with scoops of butter pecan ice cream and eat it alone in her bedroom. He said his mother's weight would fluctuate depending on the challenges she faced and ice cream she needed to self-soothe. David said she often joked about it, but no matter how little she said about the things that were troubling her at work, he knew when something was bothering his mother because ice cream helped her push away whatever feelings of stress or anxiety she might have been feeling. It was how she decompressed.

"You should come to work every day and attack the challenges before you," she said. "Walking out the door, you should be able to realize you've met the challenges of the day as well as you could, and then relax to get yourself ready for the evening and the next day's challenges. If you do not make that distinction . . . you risk walking over the edge into emotional instability."

Work-life balance was not something that was discussed much in the 1980s, and women were expected to be everywhere at once and everything to everyone who needed some little piece of them. Raye did what she could to leave work at work and home at home—and to prevent David from knowing whether she had a boyfriend or not. Although she preached knowing when to take the foot off the gas, it's easy to see how she might have struggled with that from time to time, especially given the nature of her job and the realities of her personal life. Add to that her duties on the EEO committee, and Raye was able to see firsthand what happened to people in the workplace and empathize with their struggles.

Human beings do the best they can, and the results are rarely perfect. Raye was one for likening life to a boxing ring. She said everyone gets knocked down from time to time. You should never be surprised by that, but you should always be prepared to get back up when you hit the ground.

"If you refuse to lose, then you become very difficult to defeat," she said. "On the other hand, the first time you accept defeat, you become a loser. You cannot become a loser until you decide to quit trying. If you are not really trying to be all you can be . . . you can change that. The key question is whether you will do it. It is your choice."

All her life, Raye chose to keep getting back up, no matter how hard she got hit. Later in her career, she wanted others to see that if she could get back up, they could too. By sharing her story, she was cultivating another generation of people who would share her resiliency and then nurture it in others.

15

Retirement

After twenty-three years of working with CAD/CAM technologies, Raye understood that the biggest problem was not the technology itself but in getting older employees to accept and use it. She acknowledged to a reporter that it was also challenging to integrate the technology with existing systems, but that it was worth fighting through the difficulties due to the long-term cost savings CAD/CAM provided.

In March 1988, she received the National Computer Graphics Association's Award for the Advancement of Computer Graphics at the NCGA's annual conference, because she championed the use of computer graphics in her organization. "I went out there and was fighting for this stuff when it wasn't popular," Raye said, when asked why she thought she was chosen for the award. "Other people were saying it would never fly."

In addition to her duties as acting program manager of ships, she had become the acting program manager for CAD Acquisition II, a US Navy CAD/CAM systems procurement that was said to be worth more than $1 billion. Prior to her involvement, the project was being handled, according to one undated memo "in a secretive and imperial manner. We have repaired the damage to the extent that it can be repaired. We have normalized relations with higher echelons. We have told the vendors the truth and our plans are not yet firm."

The memo went on to say that analysis of CAD/CAM Acq 2 was delayed because the correspondence files and other records had to be reconstructed by obtaining copies from other parties. "The people writing the specification lost sight of all the objectives," the memo said. "They attempted to include all equipment in any navy CAD system and nearly all of the engineering software that might be needed. Those are not proper objectives. Proper objectives must be operational objectives."

One potential data problem with CAD/CAM integration was that its service life was five to eight years, while ship data service life was twenty-five to forty-five years. So transferring digital data from one CAD equipment to another sometimes resulted in omissions and distortions that could disable whole classes of new ships.

Younger employees had become used to CAD/CAM, because they had been introduced to it in school. For them, it had become like the pencil and paper older workers had used throughout their careers. "We cannot afford to compete as a nation and have those archaic methods," Raye told a reporter. "That's why they're beating the socks off us in other nations."

Integrating CAD into existing systems was a significant expense, but it paid for itself over time, Raye said. "There's no point in printing out your drawing and then having someone else down the hall key it back in," she explained. "With good planning, you can have the thing integrated across the line. You have to design [a CAD system] with manufacturing in mind. But you have to pay the money up front in design."

The technology also allowed engineers to render several different iterations of the same design in a way that used to be costly and time-consuming. The new system allowed for better ship design because designers could experiment with different variables on the computer.

"Don't believe all the glossy ads you see, though," Raye said. "You can't just buy it and plug it in. There's more to it than that. There are some off-the-shelf systems that work, but they won't do all the fancy things that some others will."

She traced her involvement with CAD/CAM back to her CAS-DAC days, where she and others experimented with some of the concepts that would be used in CAD/CAM technology. "It all started with CASDAC in 1965," she said. "We were writing the software and trying to get the hardware to run it. We were twenty years ahead of our time." Then, the computer industry was not ready for their advances. Now, it built on those experiments and opened computer assisted design and manufacturing up to a whole new level. The navy set up the Computer-Aided Engineering Documentation System in the early 1980s, and it grew into a nearly $1 billion project for hardware and software. Raye headed a similar project, and it led to a massive navy procurement of more than $1 billion. The program was to meet the needs of all five navy systems commands. The procurement was bogged down in red tape, with additional requests for information that led to no request for proposal. All the same, Raye believed it was an excellent blueprint for the future of the navy, just one that needed to be less broad.

A colleague saw the interview in *MIS Week* and forwarded it to Raye's boss Bill Tarbell, noting that she told the reporter "no comment" three times when asked about the procurement's cost.

By the end of the year Thomas Cain, a past president of the National Computer Graphics Association, wrote to Tarbell, telling him existing specifications for integrating such a system were unrealistic and expensive to implement. "It is also unfortunate that CAD II has been publicized so aggressively throughout the planning stage as being an 'over one billion dollar' procurement," Cain wrote. "This publicity resulted in unrealistic expectations in some part of the computer and aerospace industries, and also caused unfavorable Congressional concern over the impact of the program on the navy budget. It is believed that under your direction, and with the new project management, the navy's requirements for CAD/CAM and an eventual integrated digital network will be met through a practical and orderly procurement process."

Raye had been climbing a ladder her entire career, and she wanted to continue her ascent. At this stage, it made perfect sense to her to be promoted into the senior executive service, which would have been a GS-16 job with higher pay. She was a graduate of the Federal Executive Institute, which trains high-performing employees for the SES. However, once Raye's commanding officer left, she said that they abolished the position and pay grade she wanted before she could become the navy's first Black female SES. She couldn't file a grievance for a job that was no longer there, so there was nothing she could do about it.

On January 3, 1989, there was a reorganization of the Computer Aided Design department, and many of Raye's duties had been taken from her and assigned to someone else.

She discussed the situation with her boss Bob Morgan to see what her options were. Two days later, Morgan responded with a private memo to her, in which he presented her with three options. "You could fold your tent and get an early out and retire to do the things that you have always wanted to do," Morgan wrote. "I don't recommend that option unless you really want to do something." Then, Morgan wrote she could "tread water" until she was eligible to retire. "I don't recommend that option either," he wrote. "You would get bored and not feel good about yourself." Finally, he advised that she could "aim for an area of interest that gives you something challenging to do, but doesn't blow your corporate knowledge base.... You would have to pick something that doesn't say 'CAD,' 'IRM' or any other of the political fad things." He recommended that she pursue this option because it would give her good experience when she did finally decide to leave, and she wouldn't have to be involved in the CAD and IRM efforts, which were likely to crash and burn. He closed the note with the admonition to not let the bastards get her down.

Armed with that advice, Raye went on a mission to figure out how she would spend what would clearly be her final year in the navy. Her first step was to address how she'd been mistreated in the reorganization of the CAD department: "I have worked my way up from GS-3 to GM-15 by 30 years' exceptionally diligent effort with no gift promotions," she wrote in a memo, dated January 9, 1989. "I am in fact very well qualified for the positions, and my performance ratings for the past five years include four outstandings . . . and a superior rating for the period 6/87 to 6/88. No reasonable person would believe that my sex and color did not play a role in your decision. If I were you, I would not relish the task of convincing anyone that: I have not been mistreated; that the people who signed my last five performance ratings, including you, were all lying."

Raye wrote that she did not have a serious grievance as long as her superior's actions did not reduce her pay, did not assign her to a job farther from her home than Crystal City, and did not assign her to a job with "meaningless duties unworthy of my attention."

"No one can succeed as the command's ADP leader without reasonable support from the headquarters' SESs," she wrote. "I have suffered enough during the past 32 years from being an ambitious black female. I have no desire to suffer another long fight."

She had been planning to retire in February 1990, right after she turned fifty-five and earned thirty-three-and-a-half years' worth of service. David was out of college, so the pressure she felt to work was lessened. In the meantime, she said that the command's ADP practices needed to be improved substantially.

"I did not have time to address this area personally as the deputy, acting director or director of PMS 309," she wrote. "But I am capable of making a substantial contribution in this direction before I retire."

She recommended that SEA 90C be changed to ADP Technical Practices and that she continue to be assigned to SEA 90C. With that designation in place, she said she would present a plan of action and milestones within six weeks. "I will not require any assistance to carry it out," she wrote. "I will require an office, the computer equipment now in my office, and $2,000 for software and professional books. I would like to be authorized approximately 10 days' travel to ADP conventions during the next year."

She said what she proposed badly needed to be done, and she could do it for one-third of the cost of a government contractor. She said she planned to document the best practices of in-house employees and some contractors so that other parts of NAVSEA could adopt those patterns in their own work. "It will make my last year as a gov-

ernment employee worthwhile for both the government and me," she
wrote.

It seems that the recipient of the memo did not find such work
necessary, and Raye was forced to regroup. On January 23, 1989, she
wrote another letter to her superior about her tasks and duties for that
year. She said that she met with a personnel representative about early
retirement based on discontinued service and decided that it was best
to retire after January 21, 1990. She targeted February 3, 1990, as her
retirement date. "With the idea that I now have less than a full year of
actual work time left in my formal career, I have decided that it would
be unrealistic to propose tasks and duties that could not be successfully
planned, implemented, and completed during that time," she wrote.

Then, she noted the dearth of positive role models for inner city
junior and senior high school students. "I have served in this capacity
over the past 15 years, representing the Naval Sea Systems Command
whenever I could squeeze in the time from my hectic work schedule
and most often I have provided this service on my own time," she
wrote. "Even though I tried to speak to the youth when invited—I
was never able to honor all the requests I received to address students,
church groups and civic groups because of the limitations on my time.
Normally, I encouraged minorities and women to pursue careers in
Computer Science, Naval Architecture and/or Engineering—addi-
tionally I found myself always faced with questions regarding careers
from those students who did not plan, or could not afford to attend college. Most of these young people wanted to find meaningful employment but did not know where to start or what their options were. In
situations of that nature, I have encouraged them to seek employment
with the navy while going to college at night to enhance their skills; or
once employed they have the potential of eventually obtaining training
via Upward Mobility programs, etc. This I feel is a worthwhile public
relations activity and an area where our command is not concentrat-
ing enough time or energy. This Command is desperately in need of
talented people who can be trained and encouraged to grow into true
leadership positions. If these young people are never made aware of
the potential careers we have to offer, they will not seek to prepare
themselves properly, or apply for positions when they are available."

She wrote that the command also needed more engineers in contractor support services, but that those engineers would also need support in many areas, from administrative to accounting and documentation, among other things. She wrote that she believed those requirements could be coupled with the public relations services she proposed and used as a way to advertise the navy's needs so they could better recruit personnel to provide the talent base the command needed. She recommended that her title be changed to special assistant for public relations and CSS and that she should be given the latitude to proceed as she saw fit. If given that latitude, she said she would submit a detailed plan of action and milestones within six weeks.

She said she wouldn't need a staff to do what she proposed, but she would need a private office to consult with counselors and students. She would also need to know which skills are needed among various positions so she could highlight those when she spoke.

"What I am proposing to do badly needs to be done," she continued. "I can do it, and there would not be any additional cost to the command since I am already on salary. I will document my basic presentation so that upon my retirement someone else can continue this community service while providing the command with an entry vehicle into the local schools, if it is deemed desirable. I normally use 'Navy ship design and construction' as my basic theme, then I shift my discussion to fit the interest of the audience to show them how the various careers support the navy's efforts." Raye wrote that she was capable of handling such duties well before she retired and planned to work with the superintendent of DC public schools and counselors to achieve a reasonable schedule.

For the most part, this is how Raye spent her final year with the navy. David recalled that when his mother put in her retirement papers, her boss was astonished because he thought she'd be working a while longer. "It was a demeaning conversation about how expensive college loans were," David said. "She told him that she had already paid my way through school and had money set up to send me to graduate school. She told him she had it covered and that she was going to retire."

Although Raye had had several victories throughout the course of her navy career, her final year working there reminded her of the hardships she faced and would continue to face as a woman of color. Her hopes of moving up the civilian hierarchy had proven increasingly difficult to achieve.

"Her last boss made a point of making her as miserable as possible, the closer she got to retirement," David said. "I remember the days when she would have these projects going and he would try to take things away from her. He didn't like seeing her in that role, and Mom said he kept messing with her."

By this point, David had graduated from Morehouse College and moved back to DC. "Mom would come home, throw her bags down, and just be worn out," he said. "She was so frustrated because her boss was either trying to take duties away from her, or he was taking credit for the things she was working on. Sometimes he would even say the things she did were a failure, when it took several people to do all the things she was able to do."

Near the end of her time with the navy, Raye was moved out of her office and into a cubicle. "That was a slap at her level," David said. "She had an office with a couch, a table with chairs, and all kinds of naval artwork. To be put out there with everyone else hurt her deeply. One day someone asked her if she would take back one of her duties, she told them no. She spent the last year or two doing her own thing and mentoring at different schools in the area."

She was hanging on, but she wasn't enjoying it. David always liked to come visit her at work. Before he realized that his mother had been moved out into a cubicle, he started to feel like she did not want him there. Perhaps it was wounded pride, or perhaps she wanted to keep him away from what had become a toxic environment. She went in every day to her shrunken world, where her desk was virtually empty of work, and counted the hours until she could go home.

Her one saving grace: knowing that her son was on solid footing professionally. He had been interviewed twice for a contract data analyst job with the Drug Enforcement Agency, and the talks led to a third and final panel interview.

"I visited her at her office in Crystal City, Virginia, to let her know that I got the job, provided I passed a federal government security clearance," he said. "Her eyes welled up. She understood the doors were open for me now. Growing up in Washington, DC, had its challenges, especially if you were, like me, a young Black male living in a single-parent household. The news that my foot was in the door was all she needed to know that her struggles with me were worth it." With David taken care of, Raye endured her final moments in the navy.

The day that David was to report to the FBI Academy headquarters at Quantico, Virgina, for basic training, his estranged father died. Although David did not want to attend Dave's funeral, he knew his mother was handling his late father's arrangements and that he should probably attend. Raye told him not to worry about it, to start this next chapter of his life, and that she would be fine. His mother was not one to harbor a grudge, David said, and the fact that she was burying Dave in the family plot was proof of that. David ultimately got special permission to attend the service, knowing that his mother would appreciate having him at her side. Raye believed it was important to forgive, but she also stressed that it was just as vital to never forget.

"When she would get angry, she'd be smiling a lot of the times, but other times, she'd go off on people and shout, 'You act like you're trying to screw me without a rubber!'" David said. "People would turn around and look at her, like, 'What?' So she forgave people, but she also understood that whether people acknowledged that or not, it was really her responsibility to never get in that situation again. She emphasized that with me. She said a lot of people would screw with me in life, but I had to know how to play the game."

Part of Raye's success—and what made her so memorable—was her ability to create networks of mutually supportive people. "I watched her plant seeds everywhere," David said. "When you plant a seed, something's going to grow. So when people messed with her, the reality was that she had a support network of people she had helped who looked out for her, too. And she also kept copies of everything. Just in case."

On January 31, 1990, Raye Montague was feted at a retirement luncheon attended by coworkers and extended family. A caricaturist captured her likeness. All attendees signed the rendering and presented her with all manner of gifts. Joining Raye at the head table in the banquet hall were her son David, mother Flossie, former mother-in-law Sarah Montague Crockett, aunt Gladys, former boss Captain John F. Leahy, and former coworker Art Fuller, among others. Pictures suggest it was a happy occasion, warm with affection for the woman who had touched so many of their lives in so many ways.

"Can you believe my mother lived to see me retire?" Raye said. "And it was such a surprise, but I was presented with an American flag that had been flown over the Capitol building in my honor at the request of Senator David Pryor. Can you imagine receiving an honor like that and having your mother there to see it?"

Raye was also presented with a small brass replica of a canon from the USS *Constitution*, a ship from the Revolutionary War that is the oldest commissioned ship in the US Navy.

"They only made those canons for generals and admirals when they retired," she said. "No civilians got them. But when I got a system set up at a Naval Ordnance Lab in Louisville, Kentucky, they asked me what they could do for me. And I told them to get me one of those canons when I retired. By this point they were making the canons out of alloy, but they made it out of brass just for me."

Surrounded by people who had touched her life in so many ways and reminders of her success, one would think Raye would have grown wistful or regretful that she was ending three groundbreaking decades at the navy. She had broken so many glass ceilings and so many barriers and had been the first to do so many things in so many ways. She was fifty-five years old and could have worked another decade at least. So why quit before then?

"I was tired," Raye said. "I was done."

David said his mother stayed in the house for the next few months, sleeping. She had never had free time before. Now that she did, she was going to rest up and figure out what to do with it.

16

Return to Little Rock

Raye Montague moved back to Arkansas in 2006. Her mother, Flossie, the so-called wind beneath her wings, had died of congestive heart failure in 1993, and her son, David, had become a tenured professor of criminal justice with the University of Arkansas Little Rock. By this point David had struggled through a first marriage that ended in divorce, earned a PhD in criminal justice, then remarried and become a father. He was coming into his own professionally and personally, and Raye had been retired for sixteen years. Plus, she had become a grandmother.

"Everyone knew how much she loved her granddaughter, Riley," David said. "She was so proud of her and the numerous accomplishments she had achieved in her life. She loved that Riley is constantly exposed to opportunities and encouraged. She loved that Riley is consistently made aware of her history as she learns and attempts to figure out her own place in this world. She loved that Riley is being taught to be unafraid of just being herself, and that the opinion of others does not define who she is."

Raye's legacy would transcend the way she created a new and better way of designing the navy's ships and submarines. Ships can be decommissioned, just as the Oliver Hazard Perry was in 2015. Technology changes, but there's nothing more eternal than dreams, values,

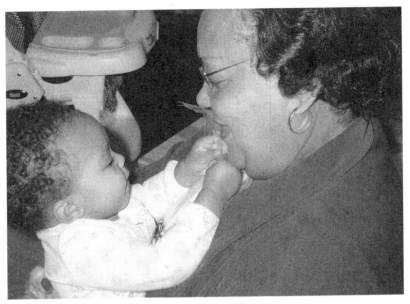

Raye with her granddaughter, Riley Grace Montague.

and personal drive, handed down through the generations. Twenty-three of Raye's Perry frigates were still active with the navies of Turkey, Egypt, Poland, Pakistan, and Spain.

Those places were far away. Raye's glories were long past, but not forgotten. There was nothing keeping her tied to the city where her life had changed beyond her wildest dreams. She wanted to start a new chapter, and she decided it needed to start near her son.

"David meant a lot to her," Sandra Howell said. "She didn't want to live far from him, so she decided to sell her house so she could live closer to him."

Raye bought a home in a gated community not far from her son's house where the homeowners preferred to sell to friends and family. Her name being Raye Montague, her job being an engineer, she assumed that her old-money neighbors-to-be would presume she was a White male. On Easter Weekend 2006, she moved into her new house, and two weeks later the women in the neighborhood threw a cocktail party so they could meet and get to know her.

"So I went and I had their caviar and all this stuff, and finally they asked me what my career was, so I told them," Raye recalled. "Well, they were awestruck, and I told them I surely wasn't there because of drug money. Then they asked me whether I had married, and I told them that of course I had. I just didn't tell them how many times. I even told them I had a son: Dr. David Montague. They asked me what I liked to do and I told them I liked to play bridge, dance, travel, and do crossword puzzles. I just went on and on like that."

A week later, the neighbors invited her to play bridge and she did better than expected. "The guy living next door to me was gay, and he always told me that I could let him know if I ran into any problems in the neighborhood," she said. "He had problems out there, being gay. So I told him I played bridge with them and he said they'd never ask me back. I said, 'That's all right. Now they know I can play bridge.'" But the neighbors did invite her back. "Maybe they recognized they could learn something," Raye said.

After growing up in a Little Rock that had vastly different racial attitudes, it's easy to see how Raye might have believed she wouldn't be accepted there after her return. But she came to accept that things had changed in the fifty years since she had been gone. "I think of the places I could not walk into where I am now received and welcomed with open arms," she said. "People hear about my achievements and assume I graduated from Central High School, but I tell them I graduated from college a year before it integrated. But I didn't achieve because of integration. I achieved because of my determination."

College buddy Bonnie Dedrick was happy to have Raye back in the state, because she had someone with whom she could visit and talk about the old times. Raye would drive down to visit Bonnie for University of Arkansas Pine Bluff's homecoming, and take part in all the activities that were going on, especially the football.

"Sometimes she would stay down here for a week or two," Dedrick said. "She was like a part of the family. She loved people, and that love was returned to her. Raye never met a stranger."

Another college friend, Lula Brooks, reconnected with Raye after she returned to Little Rock and became involved with the local Links Incorporated chapter there. Brooks was also vice president of programming for the American Association of University Women, so she decided to book Raye to speak for a Women's History Month event.

"Although she had always been like a member of the family, more or less, I didn't know too much about what Raye had been doing in Washington until she came back to Arkansas," Brooks said. "So I asked her if she would speak at this event and she told me she would. She was so good that day that it opened the door to her speaking at all these other places." Raye became a popular speaker at colleges, civic and social organizations, libraries, churches, and STEM panels and conferences. In a single week, she was the grand marshal in her alma mater's homecoming parade and inducted into the Arkansas Black Hall of Fame.

"She was a very outgoing and friendly lady who made it her business to meet people," said Boston "Baked Beans" Torrence, whose family owns a flower shop across the street from Aunt Gladys's former home. "Everybody loved her because of her disposition. Raye went way out of her way to help people."

Part of the way she did that was through her public speaking engagements. She had so many tales about her life that people told her she should write a book. But she said she never thought about it too much. She may not have felt like she had the time. After all, she was so busy reaching out and connecting with people, telling them that they could achieve, no matter their circumstances. She figured she could inspire someone who felt like the cards were stacked against them in life. She could let them know that a great support system and plenty of self-belief could take you farther than you could even imagine, as long as you put in the effort.

"When I needed to get dental implants, my dentist told me that because of CAD/CAM, he could produce the implants in under two hours," Raye said. "He didn't know he was using an iteration of a program I had debugged decades ago. Most people wouldn't want you to know they had dental implants, but I'm proud of mine. I contributed to them being produced so quickly."

A group of engineering students in British Columbia heard of Raye's accomplishments and desire to inspire others. The students were on a robotics team that builds robotic sailboats that they call "sailbots." The sailbot team sent Raye a letter in 2017 explaining how they sailed a sailbot across the Atlantic Ocean and named it after a famous engineer in Canada. In 2018, they built a new sailbot to sail from British Columbia to Maui, Hawaii, and they named that sailbot *Raye*.

One of the more poignant connections Raye made in her later years was with inmates in a reentry program at the Arkansas Department of Corrections' Pine Bluff Unit. The program was created by inmates and designed to connect them with outside volunteers with subject matter expertise in areas such as academics, the criminal justice system, housing, finance, substance abuse counseling, and spiritual development to help inmates leaving prison have a better chance of staying out. After finding out that David was a mentor in the program, Raye wanted to go to the prison to find out more about the work he was doing. She was so inspired by the initiative that she became a speaker and mentor too.

"Raye Montague is one of the most wonderful people I have ever met," said Vonnie Moore-Shabazz, the cofounder of the program, who is currently serving a life sentence. "When she told us her story and the obstacles she overcame, she gave us hope. And for her to be able to relate this to the guys in the way she did and keep them interested? This is no small thing. This place can be a tougher audience than the one at the Apollo. But she was able to relate to the guys, and show them that over time, her work spoke for itself and she became accepted among her colleagues. Look at the work she has done for this country. She is a great American. I can still see her saying, 'Who would have thought a little girl from Little Rock could do so much?' She made us believe in ourselves."

Given Raye Montague's experiences in life, both in segregated Arkansas and in the navy, two places that were not always welcoming to women or minorities, Raye said she never imagined she'd see a president of African descent. Yet, when Barack Obama became the Democratic nominee for president in 2008, Raye raised money for him, and then received an invitation to his inauguration.

"I look at that and think, when they really gave us a hard time, nobody ever dreamed this," she said. "I cried my heart out when he won that election."

As David became more and more successful, she cried tears of joy for his accomplishments, too. She was immensely proud of her son, who overcame obstacles, just as she did. "He had the benefit of a lot of things that other people his age hadn't been exposed to," she said. "And he knows, just as I did, that as you reach higher levels, you must reach back and bring others with you. That's why he has me mentoring at prisons. That's why he instills in his students the things I instilled in him."

David is raising his daughter, Riley, the way he was raised, instilling in her the belief that she will go to college and achieve. When Riley was five years old, Raye said she had a habit of poking around on computers and telling people she was working. She mimicked what she saw, and by all accounts, it's likely that she will follow in her parents' and grandparents' footsteps. As it stands, Riley tells her father that she will achieve more than he has when she grows up. She is setting her sights high.

"God, I wish everybody could have a son like David," Raye said. "He's firm, strong-willed, but kind. His students have complained about him being a taskmaster, but they've come back to him and thanked him for helping them be prepared for the outside world."

Other adults have benefitted from Raye's nurturing guidance. One of them, Donna Terrell, is a news anchor for Fox News in Little Rock. Terrell moved to Little Rock from Houston, where she had been caring for her grown daughter who had cancer. She joined the Links chapter after she moved to town, and she said many of the members, Raye included, would call and check on her to make sure everything was OK.

"Raye and I used to talk at night, because after my daughter passed away, it was very tough," Terrell said. "She was always so concerned about me, and I was grateful to have her as a friend. Sometimes she would email me after a broadcast to tell me that I was great, or that I was beaming. I remember one night I was on vacation in Punta Cana and she called me because I wasn't on the air and she was worried. I told her I was fine, and on the ocean, and she was so glad to hear it. It was one of the defining moments of our friendship. Raye was always my number one ally. Her love for me never wavered. Her concern for me never wavered. My mother passed away years ago, and Raye was a mother figure who guided me. That meant the world to me. We don't find a lot of people in our lives with the kind of knowledge that she had that can offer guidance that we need. David was so blessed to have a mother like her. She was just amazing."

Raye gave the love and care that she got growing up. In her later years, she grew concerned that children were not getting that kind of nurturing, and that it was leading to young adults who were more interested in instant gratification than hard work. She spent some of her spare time mentoring youth in the community and saw plenty of children who couldn't care less about going to school. "You have these brilliant kids and nobody spends the time to figure out what makes them tick," she said. She thought back to the career days she used to do in Washington, DC, where teachers would tell her certain students simply couldn't be reached. One of those students approached Raye, who had a model of one of her ships with her.

"When he saw that, he perked up and had all kinds of questions for me," she said. "So I explained to him what it was all about and he told me he was interested in aircraft. He had been taking cardboard from the dry cleaners and making plane models with it. I helped him get into a NASA summer program and then he went on to Massachusetts Institute of Technology. And to think he was labeled a lost cause."

She believed these problems could be overcome, but only with stronger family involvement. "We've got thirty-five-year-old grandmothers now, and babies that are raising babies," she said. "At fifteen years old, what did you know that you could teach somebody? How could you be an inspiration? Some of these grandparents are still thinking that they can enjoy their young life, but you've got two more generations that we're losing. We've got to stop that and we don't know how. Our world is out here, and we need leaders. You can't be a leader if you're out here looking like a bum and not well educated. They have to be willing to forgo some of the pleasures in life in order to achieve. Once you achieve, you can have all the goodies."

Despite her accolades from the navy, Raye's work was not acknowledged publicly until 2012, when the *Arkansas Democrat-Gazette* wrote about her. In 2016, she was recognized nationally after the publication of *Hidden Figures*, Margot Lee Shetterly's bestselling book about the Black female mathematicians at NASA who made possible some of the country's greatest achievements in space.

"If I had been accepted at the University of Arkansas at Fayette-ville, I feel now that I would have been stuck in a cubby doing mundane things," Raye said, reflecting on her life. "Because I had to go through all the obstacles and hardships, look at how the world opened up for me. Look at the lives I've touched and the lives that have touched mine. These were things I've been able to do that I never would've dreamed existed."

A local television journalist profiled Raye for a segment that garnered 2.5 million views on Facebook in the twenty-four hours after it aired. *Good Morning America* discovered the story and brought Raye from Arkansas to New York City by chauffeured limo. She appeared on GMA with Robin Roberts and *Hidden Figures* actresses Janelle Monae and Octavia Spencer.

"I want to let you know that you are no longer hidden," Spencer told Raye in a video clip. "We see you. We salute you. And we thank you."

The US Navy honored Raye as its own Hidden Figure that same year. She traveled to Washington to receive the honor, and she spoke frankly about the prejudices she faced while doing the work she loved. The NAVSEA staff applauded her and, according to David, her words brought many to tears. Raye's growing fame led to countless other speaking engagements, public appearances, and accolades. At various turns, she seemed proud of the attention and a little bit surprised that she could create such a stir. The late-life notoriety only increased her platform so that she could do what she loved.

"God gave us all many talents," she said. "If the first talent doesn't work for you, use another one and another one and another one. If someone places an obstacle in your way, drop back and take a different route. It might take you longer, but you can achieve and excel in spite of the system, not because of it. I did. You can do anything provided you are educated and work hard. You might have to run circles around other people to prove that you're where you need to be or force doors open. Otherwise, people will never know that people who look like you can do real things."

Even with travel restrictions due to her health, Raye went to New York a second time to appear on Harry Connick Jr.'s talk show *Harry*.

"My intern, Nia, went with Mom because I had a work conflict," David said. "I called to check on them that evening, and Nia told me they dealt with a little hiccup on the flight, but they were fine. They had checked into the hotel and were at an outside café drinking piña coladas. Nia told me she was just loving the opportunity to soak up so many of Mom's stories. Nia loved her energy, as so many others did."

After interviewing her on his show, Connick told Raye, "If the only conversation I've ever had is the one we just had, I feel like my life at this show is complete. It's a great, great honor to have you here."

Although Raye was able to look back at how far she had come since her earliest days with the navy, she learned that many women within that branch were still dealing with many of the same issues she faced when she retired in 1990. She believed that discriminatory practices held society back. The only way forward was to open doors for everyone, regardless of their color, gender, or beliefs.

"She was busy opening doors for people and inspiring them," David said. "Her message was always the same: 'Don't let people put obstacles in front of you, but understand you also have to put in the work.' She didn't have any patience for people who weren't willing to go the extra mile."

Raye died of congestive heart failure on October 10, 2018. Before she died, she said she would like to be remembered as an inspiration to other people. Yes, she was an engineer and a trailblazer in the navy, but she said inspiring others was the work of her life.

"I was put here for a reason," she said. "That reason is to open doors for other people."

After Raye's funeral, she was brought back to Maryland to be buried next to her mother, Flossie. One month later, she was posthumously awarded the Silas Hunt Legacy Award by the University of Arkansas at Fayetteville. Like Hunt, who paved the way for future Black students at the university, Raye was celebrated as an example of how diversity and inclusion strengthen the fabric of a community and the world.

"I loved her to death and am heartbroken about her being gone," Debra Moore-Lewis said. "She had a huge and giving heart, and she didn't judge anyone or think she was better than anyone. Just thinking about her now makes me feel emotional."

Moore-Lewis has not been alone in her grief. Since Raye's death, David has embraced the causes that were dear to her heart.* He is an engaging public speaker who often talks to school groups about the power of education, using his mother's story to captivate young minds in the way that inspires them to reach for the stars, as she did throughout her life—once even touching the moon.

---

* David Montague is doing his mother proud, between accepting posthumous awards on her behalf, speaking about her legacy, and talking to school groups about the power of achieving one's dreams. On July 17, 2019, he spoke to a STEM summer camp that was on campus at University of Arkansas Little Rock, where he works. He talked about his mother's accomplishments, read them the picture book about his mother, *The Girl with a Mind for Math*, by Julia Finley Mosca, and then gladly answered countless questions from the students. He was invited to stay for snacks but declined as he had a meeting. But if he hadn't had that meeting, Paige believes Dr. Montague absolutely would have stayed. He likes snacks, and so does she. Later in the day, the students slid several construction-paper thank-you notes under David's office door. One student wrote: "Thank you so much for coming to talk about your mother. It takes guts, especially since it has not been a year since she has passed." It does take guts, and Paige has been constantly amazed at the reserves of inner strength David has had, working a regular day job, being an engaged parent and husband, mourning the loss of his beloved mother, and working with a nerd in Atlanta to make this book become real.

ACKNOWLEDGMENTS

Many authors talk about the lonely process of writing a book. For more than a year, we have been fortunate to have each other to lean on and laugh with as we've made these pages come to life.

In Little Rock, we would like to thank the staff of the Center for Arkansas History and Culture, a division of the University of Arkansas at Little Rock, for allowing us to come in on short notice to look through some of the items in Ms. Montague's private holdings there. We appreciate their commitment to preserving private papers and other ephemera from her long, rich life. Another big debt of gratitude goes out to Patrick George Williams at the University of Arkansas, and Travis Ratermann at the Arkansas Historic Preservation Program, both of whom were able to determine that the provenance of the submarine Raye saw in her youth was Japanese, not German.

Raye was a wonderful and feisty storyteller, which is little wonder, as she lived a wonderful, feisty life. We were fortunate to get what memories we could from her while she was still alive, and those tales form the backbone of this book. Many people graciously stepped up to fill in what blanks there may have been, and we are grateful to them for mining their memories. In no particular order, we would like to thank Bonnie Dedrick, Marge Coleman, Rosenwald Altheimer, Art Fuller, Peter Bono, Trenita Russell, Sandra Howell, Larry Howell, Boston "Baked Beans" Torrence, Donna Terrell, George Brown, Vonnie Moore-Shabazz, Debra Moore-Lewis, and Lula Brooks. All of them have provided us with valuable stories from Raye's life as a young woman, new graduate, young mother, up-and-coming civilian engineer, trailblazer, neighbor, friend, mentor, and retiree enjoying her final years.

We are grateful to Jerome Pohlen, our editor at Chicago Review Press, who came from engineers and wanted us to tell a story about one. He has been kind, patient, supportive, funny, and wonderful to work with. Bottomless thanks to him for championing Raye, and, in turn, championing us.

Speaking of champions, we are so very thankful for our agents Jane Dystel and Miriam Goderich at Dystel, Goderich & Bourret, LLC. They made all of this possible in so many ways and throughout various iterations, changes, ups, downs, curveballs, and the like. Everyone should have an agent who is willing to commit the time and passion to making sure inspiring stories like this not only get discovered, but see the light of day. Their belief in and support of us, for better or for worse, has meant more to us than they'll ever know.

From David

It is such an honor to have had this opportunity to join in the telling of my mom's story. She started this project as a memoir and was working with Paige Bowers, my amazing partner on this project. When it became apparent that Mom could not continue, I promised her I would make sure her project was completed. So much has come together to make this book happen, and the experience has given me an even more wonderful appreciation for the contributions and strength of the late Raye J. Montague, my mom.

My family and friends have been with me nonstop and encouraged me with their kindness and concern as I've handled this transition since October 10, 2018. Notably, my wife Whitney and daughter, Riley, have pushed me to do whatever is needed to find time and energy for this undertaking. My friends have also supported my efforts to stay focused on this: namely, Scott Webb; Larry Howell; Eric Walker; Artemesia Stanberry; and Pearl, Dowe, and Becky Alred. Without a doubt, Paola Cavallari of the University of Arkansas Little Rock History Department and the Clinton School of Public Service played a key role in creating a significant index of hundreds of items from my files and piles of loose materials, aiding this book project. I also appreciate the significant support of coworkers Donna-Rae Eldridge, Moyosooreo-luwa Kemi-Rotimi, Pradeep Mahalingam, Antonia (Nia) Billups, and Kristen Peterson in so many ways.

I also cannot express how much the suggestions from Mom's sorority sisters in the Beta Pi Omega Chapter of Alpha Kappa Alpha Sorority Inc have been; they provided additional understanding on the cultural importance of her being an AKA, as it relates to her passion for helping others. Also, the Little Rock Chapter of the Links Inc. and the Prince George's Chapter of the Links Inc. played a valuable role in providing information about Mom's long years of service via that sisterhood.

Also, to me, this book that Mom wanted to create as an inspiration to others, was heavily supported by all those who served with the military and by those who helped to educate others. Your expressions of support and your suggestions on important military and educational nuances also aided this book. Without a doubt, I am so thankful for being able to participate in numerous tributes to my mother over the past few years in schools, churches, prisons, youth summer camps, etc. Those tributes, some of which included reenactments, scholarship ceremonies, and others with me sharing her message with others, gave me an even deeper understanding impact of Mom's impact and legacy.

Finally, Paige Bowers came into our lives, and once Mom and I met her, we knew she was the perfect partner on this project. Paige was there to give this project the kick start it needed. She was always a friend I was able to call (or text) during the difficult time of transition. We've worked so well together covering research and ironing out details, and while she is an amazing writer, I am stumped as to how she has learned so much about my family in such a short time. I know that our connection will continue far beyond this project.

From Paige
I am very fortunate to be surrounded by a strong, loving support system. My friends, family, and colleagues have been just as eager to get me out of my head and away from my desk as they have been to cheer me on as I sit at it, talking on the phone and typing away. Many, many thanks to: Colleen Orange Adams, Julie Baggenstoss, Kathy and Matt Bedette, Andrea Billups, Regina H. Boone, the Diecks family,

the Daflers, Aly Fickenscher, Sheryl Bagsby Frison, Fiona Gibb, Amy Haimerl, Michelle Havich, T. Sidney Hollwager, Theresa Kaminski, Alison Law, Bekah Kendrick, Kara McDaniel, Kim MacLeod, Ryan Mikolosik (told you I wouldn't forget this time!), Daphne Nikolopoulos, the Ortego family, the Parr family, Carolyn Porter, Jill Rothenberg, Elizabeth Rynecki, Nikole Sarvay, Kenna Simmons, Pamela D. Toler, D. J. VanCronkhite, Katherine and Tony Warren, and the Veit family.

My mother, Pat Bowers, has been an incredible sounding board, as she was once a single mom working for a navy contractor right around the same time Raye Montague was with the US Navy. Her sense of the culture and the moment was invaluable to me, as were her recollections and what she, too, did to excel in a man's world and raise two children too.

My husband, Jeffrey Diecks, has been, as always, an amazingly supportive and loving partner throughout this. I've said it once, but I'll say it again: I seriously do not know what I would do without him. My daughter, Avery, has blossomed into a high schooler before my very eyes. Seeing her grow into a bright and promising young woman has been one of life's greatest gifts, in a life that already has so many of them. There is also Murray, my dear dog, who many people mentioned here have heard barking in the background as I've interviewed them. He still sits at my feet or on them as I write, and it's because he is a very good boy.

Last but not least, I need to thank the Montague family for trusting me with a treasure. It has been a great honor to tell this story about a woman who simply would not take no for an answer, and I've never taken the privilege of doing this lightly. Raye Montague inspired countless people throughout her life, myself included, and I am grateful to have been offered a window into her world.

As for David Montague, we have been on quite a journey, you and I, from the first day we talked to each other about helping your mom write a book. Who could have imagined what would have unfolded since that first day? I am grateful to you, my book partner, and am proud to call you a friend and brother from another mother. I know your mom has been smiling down on us as we've turned one word into several thousand on her behalf. No matter what name goes on the dedication page, this book is, without a doubt, for you.

HONORS, ACCOLADES, AND ACCOMPLISHMENTS

2018
Silas Hunt Legacy Award (Pioneer Award), posthumously awarded
Arkansas Women's Hall of Fame, inductee
Honorary Doctor of Laws degree, University of Arkansas at Pine Bluff

2017
Raye Montague Scholarship for a "Black Woman Engineering Student" established, Oklahoma State University, Stillwater, Oklahoma
Citation, Arkansas House of Representatives
Citation, Arkansas Senate
Lifetime Achievement Award, University of Arkansas Pine Bluff, Pulaski County Chapter
AKA Founders Day declared "Raye J. Montague Day," Pine Bluff, Arkansas
"Trailblazers Award," first recipient, University of Arkansas Little Rock Bowen School of Law, Black Law Students Association, Little Rock, Arkansas
Recognized by the US Navy as its "Hidden Figure" at Naval Sea Systems Command (NAVSEA) Headquarters, Washington Navy Yard, Washington, DC
Recognized by the US Navy for lifetime contributions to naval innovation during an event at the Naval Sea Systems Command research facility, Dahlgren, Virginia

Raye and David when she was selected as a Spark for the Museum of Discovery in Little Rock in 2017. A Spark is an Arkansan who has had success in science, technology, and math after intensive study.

2016

Received Platinum Link Status, the highest level within the Links Inc.

2013

Arkansas Black Hall of Fame, inductee

2012

Certificate for Service, University of Arkansas Little Rock Re-Entry to Society Program for inmates at the Pine Bluff Prison Unit, Pine Bluff, Arkansas

2010

"Worthy of a Nobel Nomination," awarded by Hampton University, Hampton, Virginia

1990

Flag flown over the nation's Capitol in honor of retirement from the Naval Sea Systems Command, presented by Senator David Pryor

Resolution in honor of retirement from the US Navy, Maryland State Legislature.

1988

Award "For the Advancement of Computer Graphics," National Computer Graphics Association, presented by Mr. Roy Edward Disney of the Walt Disney Company

1986

Award for assistance establishing a campus Navy ROTC Unit, Morehouse College

Award for volunteer work on behalf of disadvantaged youth, Superintendent of Public Schools, Washington, DC

Distinguished Alumnus of the Year Honor, National Association for Higher Education, University of Arkansas, Pine Bluff

1985

Letter of Commendation, Office of Personnel Management

Letter of Commendation, Department of Commerce

Letter of Commendation, Asst. Secretary of the US Navy

Distinguished Alumni Citation of the Year Award, National Association for Equal Opportunity in Higher Education (NAFEO)

1981

Award as a "Career Day Speaker" at Dunbar High School, Washington, DC

Award for Distinguished Service on the Initial Graphics Exchange Specification (IGES) steering committee, National Bureau of Standards.

1980

Kellogg Lecturer, University of Arkansas, Pine Bluff

Arkansas Traveler Award, presented by Governor Bill Clinton

Honorary Citizen of Little Rock, Arkansas, presented by Mayor Lottie Shackleford

Outstanding Performance Award and Special Achievement Award, NAVSEA

Award from Computer and Automated Systems Association (CASA) of the Society of Manufacturing Engineers (SME) "in recognition of substantial contribution to the association"

1979

Award for "Outstanding Service and Leadership" on the executive board of directors, Numerical Control Society

Award for service as the federal government's only participant on the long-range planning committee for the AutoFact II Conference, Society of Manufacturing Engineers (SME)

Guest Lecturer on CAD/CAM, US Naval Academy, Annapolis, Maryland

Registered and granted designation as a Registered Professional Engineer in the area of Manufacturing Engineering, Canadian Council of Professional Engineers Certification

Reelected national officer on executive board of directors, International Numerical Control Society

Manufacturing Engineering Achievement Award "with the distinction of being the first female professional engineer in the United States to receive this award," Society of Manufacturing Engineers (ISME)

1978

Elected a senior member, Society of Manufacturing Engineers

Registered as a Professional Engineer in the area of Manufacturing Engineering, California State Board of Registration for Professional Engineers

Letter of Commendation, Secretary of the US Navy

Letter of Commendation, Chief of Naval Material

Elected International Officer on the Executive board of directors, International Numerical Control Society

1977

Letter of Commendation, Chief of Naval Material

Letter of Commendation, Department of the Air Force

1976

Letter of Commendation, Assistant Secretary of Defense

1975

Appointed head of the Computer Aided Design Numerical Control (CDNC) Lead Service, Department of Defense

1974

Represented Department of Defense and the United States on the international panel, "Ship Design and Construction via CAD/CAM and Numerical Control," Toronto, Canada

Published journal article, "Computer-Aided Ship Design and Construction: The Navy's Approach to CAD/CAM," *Journal of Numerical Control*, July 1974

Presented paper "CASDAC, The Navy's Approach to CAD/CAM," International Numerical Control Conference

1973

Selected as Federal Women's Program Coordinator

Letter of Commendation, Chief of Naval Material

Federal Woman of the Year Award nominee, Department of the Navy

1972

Outstanding Performance and Special Achievement Awards NAVSEC

Navy Meritorious Civilian Service Award

1971

Credited with the rough draft of the first US Naval ship design using a computer

1956

BS in Business, University of Arkansas AM&N (now the University of Arkansas at Pine Bluff)

1952

Merrill High School graduate, Pine Bluff, Arkansas

NOTES

1: Little Girl from Little Rock

Many of the details in this chapter—and throughout this book—are gleaned from Raye Montague's memories, shared in various interviews throughout the course of researching this work while she was still alive.

Raye could type and was well trained. Her Application for Federal Employment (Standard Form 57 with the US Civil Service Commission) shows that she spent four months working as a clerk-typist for the Dean of Instruction at AM&N, per a class requirement. She was a teaching intern for two months after that, during which she claims to have been "responsible for all activities of the office when the principal was away."

Additional details about Montague's job interview with the David Taylor Model Basin were provided to Paige Bowers by Marge Coleman in an interview dated January 14, 2019.

For the history of the Mississippi River Flood of 1927's impact on Arkansas, we relied on "Flood of 1927" from encyclopediaofarkansas.net, "1927 Flood Changed Arkansas, U.S." from arkansasonline.com, "The Great Arkansas Flood of 1927" from www.geology.arkansas.gov/docs/pdf/geohazards /1927Flood.pdf, and "Mississippi River Flood of 1927" from Britannica.com.

"one of the devil's great tricks" is a line in Charles Baudelaire's posthumously published poetry collection, *Paris Spleen* (New York: New Directions Publishing, 1970). In it, Baudelaire laments about the struggles of the poor and about inequalities in modern life.

"prowled about in broad daylight, looking for victims to devour," is a reference to 1 Peter 5:8: "Your adversary, the devil, prowls around like a roaring lion, seeking someone to devour."

For information about "Wild Red" Berry, we turned to "Wild Red Berry Was the Mouth That Roared" from Slam Wrestling: http://slam.canoe.com/Slam /Wrestling/2010/05/31/14197786.html.

For background on racial tensions in Arkansas after the flood, we turned to "John Carter (Lynching Of)" at https://encyclopediaofarkansas.net

/entries/john-carter-2289; "John Carter: A Scapegoat for Anger" at https://
abhmuseum.org/the-lynching-of-john-carter, "Little Rock's Last Lynching
Was in 1927, but the Terrible Memories Linger," in https://arktimes.com
/news/cover-stories/2000/08/04/little-rocks-last-lynching-was-in-1927
-but-the-terrible-memories-linger, and "Sandwiching in History: First
Presbyterian Church" in http://www.arkansaspreservation.com

For a fascinating look at the history of West Ninth Street, please see the
documentary "Dream Land: Little Rock's West 9th Street" at www.aetn.org
/programs/dreamland, as well as *Temple of Dreams: Taborian Hall and Its
Dreamland Ballroom* (La Vergne: Lightning Source, 2012) and *End of the
Line: A History of Little Rock's West Ninth Street* (Little Rock: Center for
Arkansas Studies, 2003), both by Berna J. Love.

2: The Submarine

We traveled a world away from Little Rock in this chapter, at least in research,
because Raye had made a captured submarine a part of her origin story. She
had long believed that the mini sub she saw as a seven-year-old was German,
and that it had been captured off the coast of the Carolinas. Given the timing
of her story, our research showed that that was unlikely. German submarines
weren't captured off the Carolina coast until a few years later. Some of the
articles we reviewed about this include: "U-boats Off the Outer Banks" by
Kevin P. Duffus at www.ncpedia.org/history/20th-Century/wwii-uboats and
"The Last German U-Boat Captain Dies at 105" by Bruce Henderson at
www.charlotteobserver.com/news/local/article213455859.html.

Memory is a tricky thing, especially when it involves a seventy-five-
year-old detail. Patrick George Williams with the University of Arkansas
and Travis Ratermann with the Arkansas Historic Preservation Program
were able to help us discover not only that the sub was Japanese, but it was
involved with the Pearl Harbor attacks, and then captured and used to raise
war bonds on a tour that snaked through the United States and wound up in
Little Rock at around the time Raye remembers touring such a sub. Today,
the HA-19 sub Raye saw in her youth is on display at the National Museum
of the Pacific War in Fredericksburg, Texas. Aside from Williams's and
Ratermann's help, we consulted the Pearl Harbor website (visitpearlharbor.
org) for additional details about the attacks. "The Midget Subs That Beat
the Planes to Pearl Harbor," by Christopher Klein, December 6, 2016, for
history.coml; Craig Nelson's *Pearl Harbor: From Infamy to Greatness* (London:
Weidenfeld & Nicholson, 2016); and Gordon W. Prange's *At Dawn We Slept:
The Untold Story of Pearl Harbor* (New York: Penguin, 1982) were also helpful
background.

The Encyclopedia of Arkansas provided us with initial information about the
Japanese incarceration camps in the state. We also consulted John Howard's
Concentration Camps on the Homefront (Chicago: University of Chicago
Press, 2008), and George Takei's *To the Stars: The Autobiography of George*

Takei (New York: Gallery Books, 2015). Cheryl Greenberg's article "Black and Jewish Responses to Japanese Internment" in *Journal of American Ethnic History* (Winter, 1995; pp 3–37) was also a valuable resource, as was William Cary Anderson's "Early Reaction of Arkansas to the Relocation of Japanese in the State" in the *Arkansas Historical Quarterly* (Autumn, 1964), pp 195–211. Finally, the National Archives has extensive holdings on Japanese internment camps during World War II. We consulted some of those holdings for our own background.

For information about Black soldiers in World War II and the injustice of fighting for rights overseas that they did not have at home, we turned to articles such as "African American GIs of World War II: Fighting for Democracy at Home and Abroad" (www.militarytimes.com/military-honor /black-military-history/2018/01/30/african-american-gis-of-wwii-fighting -for-democracy-abroad-and-at-home) and "Why African American Soldiers Saw World War II as a Two-Front Battle" (www.smithsonianmag.com /history/why-african-american-soldiers-saw-world-war-ii-two-front -battle-180964616). Other valuable resources include Linda Hervieux's *Forgotten: The Untold Story of D-Day's Black Heroes at Home and Abroad* (New York: Harper, 2015) and Gail Lumet Buckley's *American Patriots: The Story of Blacks in the Military from the Revolution to Desert Storm* (New York: Random House, 2001).

For information about the murder of Sgt. Thomas Foster, we turned to the documentary *Dream Land: Little Rock's West 9th Street* from aetn.org /program/dreamland, and "City Patrolman Shoots Negro Soldier, Body Riddled While Lying on Ground." *Arkansas State Press*, March 27, 1942, 1.

We referred to www.eeoc.gov/eeoc/history/35th/thelaw/eo-8802.html for information on FDR's attempts to prevent racial discrimination in the armed forces via Executive Order 8802.

On voting rights, we consulted "The Long Fight for the Vote: On the 50th Anniversary of the Voting Rights Act, a Brief History of African-American Enfranchisement and Disenfranchisement," *Arkansas Times*, February 4, 2015.

In addition to the assistance we received about the HA-19's background, we consulted the following news clippings for information about its tour through Arkansas: "Captured Jap Submarine Will Be Shown in Hope Nov. 19 to Aid the Sale of War Bonds," *Hope Star*, October 28, 1943; "Midget Japanese Submarine," (Fayetteville) *Arkansas Democrat Gazette*, November 2, 1943; and "Jap Sub Begins Last Half of Tour," *Hope Star*, November 15, 1943. In this article, it notes that the early November stop in Little Rock netted $15,306.60 in war bonds and stamps: "Jap Submarine, Which Is Being Shown Here Tonight; Admission by Purchase War Bonds," *Hope Star*, November 19, 1943. Further background and detail was gleaned from "$1,124 Netted War by Visit of Submarine," Hope Star, November 20, 1943.

3: Life in Pine Bluff

Moving to Pine Bluff at a tender age was hard on Raye, compounded by the fact that she had a stepfather that she didn't like. For information about Pine Bluff, we traveled through the town firsthand, relied on Montague's and Bonnie Dedrick's memories, and consulted the *Encyclopedia of Arkansas* entry on philanthropist Joseph Merrill.

For information about integration of schools, we turned to Roland Smith's thesis "Attitudes Toward Public School Integration in Arkansas Before 1954" submitted to the Faculty of Atlanta University in Partial Fulfillment of the Degree of Master of Arts, June 1961, which is an extensive look at how segregation impacted Black students and teachers in the state, rich with related studies that we mention in this chapter.

For information about Silas Hunt, we referred to the *Encyclopedia of Arkansas* entry about him, and the University of Arkansas Special Collections Entry about his life and holdings there, which include his law school exam books.

For information about the Fair Employment Practices Committee and postwar attempts to achieve civil rights for all, among the books we consulted were: David M. Kennedy, *Freedom from Fear: The American People in Depression and War, 1929–1945* (New York: Oxford University Press, 1999); Allida Black, *Casting Her Own Shadow: Eleanor Roosevelt and the Shaping of Postwar Liberalism* (New York: Columbia University Press, 1996).

Robert Penn Warren traveled throughout the South conducting interviews with civil rights icons in the 1950s and 1960s. His effort resulted in the book *Free All Along* (New York: New Press, 2019), which was the source of the interview with psychologist and social activist Kenneth B. Clark.

After the war, Harry Truman was concerned about the "moral dry rot" of racism and he attempted to address it. He ultimately found, however, that most white Americans were unwilling to confront the issue, preferring to settle in and enjoy the post-war peace. We consulted David McCullough's biography *Truman* (New York: Simon & Schuster, 1992) to gain a greater understanding of his hopes for a racially equitable country.

Marge Coleman spoke with David Montague about how Black families took to train travel, although many of them were only able to afford the ride itself. Their conversation took place on July 29, 2019.

4: Aiming for the Stars

For information about the effort to draft Dwight D. Eisenhower as the Republican presidential nominee in 1952, we consulted David Halberstam's *The Fifties* (New York: Villard Books, 1993).

Raye's senior memories album and graduation program provided details used in the narrative about her high school graduation.

For information about the Manhattan Project, postwar politics, and bomb development, we consulted McCullough, *Truman*, and Halberstam, *The Fifties*.

Computers arguably played a crucial role as nations began developing more destructive bombs and modernizing their militaries in general. To explain this, we consulted Halberstam, *The Fifties*.

Women stepped into this new world of computing and played a crucial role in handling these new, bulky machines. For information on this, we consulted: Janet Abbate, *Recoding Gender: Women's Changing Participation in Computing*; Kurt Beyer, *Grace Hopper and the Invention of the Information Age*; W. Barkley Fritz, "The Women of ENIAC," in *IEEE: The Annals of the History of Computing* (Vol. 18, No. 3); and Margot Lee Shetterly, *Hidden Figures: The American Dream and the Untold Story of the Black Women Mathematicians Who Helped Win the Space Race.*

Bonnie Dedrick spoke with Paige Bowers about meeting Raye in college on January 15, 2019.

Raye's college transcripts and notes about the subjects she was qualified to teach come from Raye J. Montague's private archive in Little Rock. At the time this book was being written, her papers had yet to be processed by the Center for Arkansas History and Culture.

Rosenwald Altheimer spoke with Paige Bowers on April 11, 2019.

David Montague spoke with Paige Bowers on September 20, 2019, about how his mother had no idea how she would have paid for AM&N if it hadn't been for that truck accident that broke her leg.

Raye's AKA cardigan, which is made of thick white and green wool, is currently held at the Center for Arkansas History and Culture.

For information on *Brown vs. Board of Education* and its impact on Arkansas, among other things, we turned to the aforementioned Smith, "Attitudes Toward Public School Integration" and Melba Patillo Beals, *Warriors Don't Cry: The Searing Memoir of the Battle to Integrate Little Rock's Central High* (New York: Simon & Schuster, 1994).

Marge Coleman spoke with David Montague on July 29, 2019, about being Grand Central Station for people trying to get on their feet and about hosting her brother and Raye's wedding at her house.

5: Exodus

Marge Coleman spoke with Paige Bowers on January 14, 2019, about the end of her brother's marriage to Raye and about her efforts to help Raye find a job.

One of Raye's applications for federal employment confirmed her clerk-typist duties. While the document itself is undated, it shows that she was looking for a better job, another raise (she had already had one that increased her salary from $3,175 a year to $3,260 a year), and elevation in government pay grade status. Comparing this document to others in her files, it seems like she sought this third raise in October 1957, a little more than a year after she began working at David Taylor Model Basin. Successive documents illustrate two more promotions over the course of the next year because of her work on the UNIVAC, as well as a constant desire to keep her computing knowledge and administrative skills as fresh as possible.

We consulted the following legal documents for information about Raye and Weldon's divorce: District of Columbia Court of General Sessions Domestic Relations Branch, Civil Action D-1063-62, which dissolved Raye and Weldon's marriage on September 29, 1965; on the same date, the Municipal Court for the

District of Columbia Domestic Relations Branch ruled that Raye and Weldon should jointly divide the proceeds from the sale of the real estate they owned.

Raye kept a file about Weldon Means until the end of her life. In it, we were able to consult documents such as: Weldon's lease of a 1954 Ford from the Capitol Cab Cooperative Association Inc., dated August 24, 1960; a letter from attorney Thurman Johnson to the Credit Bureau about Weldon's debts in Arkansas preventing Raye from obtaining credit, dated December 12, 1962; Weldon's eligibility for VA Benefits due to his service in World War II; and a letter from Weaver Bros. Mortgage Bankers, Inc., dated April 17, 1957, that said Weldon's VA benefits enabled them to process his home loan.

6: Making Waves in the Navy

Raye kept extensive personnel papers, many of them duplicates, because she knew the importance of advocating for herself in the workplace and providing paper proof of her experience, skills, responsibilities, training, and pay in the process. Those papers demonstrate that she was a woman who, while she may have struggled with certain coworkers, was still excelling and getting raises and promotions on a fairly regular basis. We turned to an October 30, 1963, Experience and Qualifications Statement (Federal Standard Form 58) for a glimpse of what she was doing in the workplace at the beginning of this chapter.

For information on the LARC computer, we consulted J. P. Eckert's paper, "UNIVAC- LARC, The Next Step in Computer Design," which was presented at the AIEE-IRE's joint computer conference, December 10–12, 1956.

There was a fascinating—and sometimes frustrating—dynamic between President John F. Kennedy and Rev. Martin Luther King Jr. during the civil rights movement. Because Raye looked up to both men, we wanted to explore their relationship and its impact on the times. We read Steven Levingston, *Kennedy and King: The President, the Pastor, and the Battle Over Civil Rights* (New York: Hachette, 2017). The media also played a pivotal role in highlighting the struggles of the moment, and the deaths of these two towering figures. For background, we consulted Gene Roberts and Hank Klibanoff, *The Race Beat: The Press, the Civil Rights Struggle, and the Awakening of a Nation* (New York: Knopf, 2006). We also double-checked a few dates and events on the University of Virginia's Miller Center website: https://millercenter.org/president/john-f-kennedy/key-events.

David Montague spoke with Paige Bowers about his mother's thoughts on the civil rights movement on September 20, 2019.

We consulted Stanford University's Martin Luther King Jr. Research and Education Institute website for King's Statement on the Kennedy Assassination. His remarks can be found here: https://kinginstitute.stanford .edu/king-papers/documents/statement-john-f-kennedy-assassination.

On July 20, 1966, Raye filed for maternity leave that began August 1, 1966, and ended November 11, 1966. Her paperwork for leave was in her personnel papers. According to a Notice of Personnel Action document, she returned to duty November 14, 1966.

The birth announcement for David Ray Montague is held in the Raye J. Montague private archive.

Rayford Jordan was not the only member of his family to sue for the right to their family land. According to court documents in the Raye J. Montague private archive, three other relatives joined him in an effort to get land from the Linnie J. Taylor Nunn estate. Each of the people named in the documents got 3/20 of the farm. On January 13, 1969, the surveyor described the plot Raye obtained as "commence on an iron stake, being SW corner of NE ¼ of SE ¼ of section 31, Township 14, Range 18. Run west for 180.5' to an iron stake at point of beginning of this description. Run south for 1354' to center of road, at a point 180.5' west of a fence, thence west along center of road for 130', thence North for 1354', thence west for 402' to a fence, then N OO degrees, 44'W along fence for 892' to a point being 76.5' South of corner of fence, then East for 541', thence south for 892' to the point of beginning of this description. All being in the West ½ of the SE ¼ of Section 31, Township 14, Range 19, Noxubee County, Mississippi. Containing 15 acres more or less." The surveyor also included an ink sketch of the plot he described.

7: A Change Is Gonna Come

David Montague had multiple conversations with Paige Bowers about his mother's complicated relationship with his father. His thoughts on closure come from the conversation they had on July 19, 2019.

David spoke with Paige Bowers about moving to Hyattsville on September 27, 2019.

Paige interviewed Debra Moore-Lewis on May 7, 2019.

Letter from Karl M. Dollak to Dave Montague, dated September 30, 1970. In a letter dated September 17, 1970, obstetrician Howard D. Wood, MD, wrote a letter that said Raye would be admitted to the hospital for major surgery on October 1, 1970, and that she needed six weeks to recuperate. David Montague talked to Paige Bowers about his mother's health, confirming that the surgery was a hysterectomy, and about Dave's rehab stint on October 23, 2019.

We turned to the following sources for information about the CASDAC program: "Introducing CASDAC: Computer-Aided Ship Design and Construction," *All Hands Magazine*, January 1974, pp 8–13; and "Development of Program for Computer-Aided Structural Detailing of Ships," US General Accounting Office, July 19, 1971.

For a primer on shipbuilding, we started with Britannica Online's article about Naval Architecture (www.britannica.com/print/article/406846), then referred to Thomas C. Gillmer, *Modern Ship Design* (Annapolis: United States Naval Institute, 1970). Peter Bono spoke with Paige Bowers about the finer points of shipbuilding on February 6, 2019, and she has since become convinced that it is one of the highest and finest arts. Much gratitude to him for his patience with an interviewer who faced an incredibly steep learning curve.

Art Fuller spoke with Paige Bowers on February 12, 2019.

The undated letter to W. Dietrich, referred to in this chapter, was written on lined paper in Raye Montague's sweeping, schoolgirl cursive. It is in the Raye J. Montague private archive.

8: Impossible Tasks

For the state of navy shipbuilding at this time, we consulted Raye's notes and speeches, as well as "Carderock: Ship Research and Development," *All Hands*, May 1971, www.navy.mil/ah_online/archpdf/ah197105.pdf.

Art Fuller spoke with Paige Bowers on February 12, 2019.

We consulted Raye J. Montague's private archive for information on the Ship Specifications program.

Raye Montague's Travel Voucher #F9855 documents the amount of time she spent in New York City working at M. Rosenblatt and Sons.

For information on the Oliver Hazard Perry frigate, we consulted the following Destroyer History Foundation article: http://destroyerhistory.org/coldwar /oliverhazardperryclass.

Letter from K. E. Wilson to Raye Montague, dated November 2, 1972, Raye J. Montague private archive.

Nomination letter for Federal Woman's Award written by Wallace Dietrich, Raye J. Montague private archive.

9: Equal Opportunities

David Montague spoke with Paige Bowers about his mother's desire for work-life balance and for being present in his life on July 19, 2019.

David's first grade photo from Our Lady of Sorrows indicates that he was one of three Black children in a class of twenty-five. Of the three Black students, he was the only boy.

Paige Bowers spoke with Peter Bono on February 6, 2019.

Paige Bowers spoke with Sandra Howell on January 14, 2019.

Paige Bowers spoke with Larry Howell on February 12, 2019.

The navy realized it had a problem with racism in the 1970s. We consulted copies of *All Hands* from that era to help us explain what was happening at that time.

John Wayne's presence at the Oliver Hazard Perry ship commissioning ceremony was a tidbit too interesting to pass up in our narrative. We consulted NavalTheater.com (https://navaltheater.com/ships/uss-oliver -hazard-perry-ffg-7) for details on how Wayne played a swashbuckling role in this ceremony.

10: Love and Happiness

According to their marriage license, Raye and James Parrott got married on May 11, 1973.

The depositions, taken in Prince George's County, portray a heartbreaking, ugly divorce in progress. Both sides were eager to hold on to whatever assets they had going into the relationship, and to take whatever else they could get from the other before their marriage was dissolved. James didn't do himself any favors as he was being deposed; his admitted "equilibrium" problems became a way for attorneys to call some of his actions and recollections into question. Raye, for her part, comes across as very earnest, very conscientiously innocent. But when it comes to talk of her assets, she hardens. It is clear that she was going to give up nothing to James Parrott.

Paige Bowers spoke with George Brown on April 6, 2019.

Paige Bowers spoke with David Montague on July 19, 2019, about some of the challenges he faced growing up, going to school, and dating. Paige felt the spat Raye had with the mother of one of David's girlfriends was too good to leave out, because it showed how fiercely protective she was of her son. Paige and David also had a good laugh talking about how awkward Flossie felt as she heard some of the things coming out of her daughter's mouth that night.

11: Another Direction

Raye received her engineering certification from the California State Board of Registration for Professional Engineers on August 16, 1978, and from the Canadian Council of Professional Engineers on July 20, 1979.

The speech Raye made at the Naval Academy was dated April 11, 1979.

Raye's notes "Requirements for an Outstanding for 1980" are dated April 19, 1979, and held in the Raye J. Montague private archive. She had received outstanding performance ratings from acting commanders in the early seventies for her work, which "consistently exceeded the normal requirements of your job and resulted in many significant contributions." Among the letters from this period attesting to her outstanding performance is "Outstanding Performance Rating Justification" from Commander, Naval Ship Engineering Center, to Raye J. Parrott, April 16, 1974, from the Raye J. Montague private archive.

A sample of the letters that requested Raye to speak for various groups includes Mae F. Wilson to Rear Admiral W. C. Barnes, US Navy, November 1, 1974. Wilson, a math teacher, was one of Raye's neighbors, and she wanted her to come speak to her class for Career Education Week. In the letter, she says, "The Mathematics Department at Dunbar Senior High School has as its prime object career awareness for all students on all grade levels. As a part of this, we are particularly interested in Federal programs and opportunities for minorities and women."

Wally Dietrich's obituary ran April 8, 1980. It is unknown which paper it was published in, but it was likely one of the main dailies in Washington and/or Baltimore.

Letter from Elaine D. Simons to Rear Admiral James W. Lisanby about Raye Parrott's school visit, April 19, 1980; Letter from Lisanby to Raye J. Parrott with letter from Simons, April 25, 1980, Raye J. Montague private archive.

David Montague spoke with Paige Bowers about his mother's interest in psychics on October 21, 2019. Raye's five pages of notes from a spring 1981 session with Jerome Groom is in the Raye J. Montague private archive.

The clipping about the Duke University study was undated, and found in the Raye J. Montague private archive.

The CASDAC Operational Environment Study is in the Raye J. Montague private archive.

The challenges Raye faced with her boss Ray Ramsay were spelled out in agonizing detail in formal paperwork and handwritten notes from Raye's personal papers. After the glories she had under Wally's tutelage, these papers show a man who is systematically and unapologetically trying to destroy Raye's good reputation at work. His comments about how personality is not enough to have someone excel in Raye's position are belittling and a slap for someone with her expertise, who had accomplished what she had after decades of hard work. Throughout his complaints and evaluations, Ramsay is demeaning and vicious. David Montague said it was treatment like this that often forced his mother to keep impeccable paper trails of everything that happened to her in the workplace, and to forge alliances with people who would protect her, or at least vouch for her, in the face of such treatment, as her supervisor Jerry Cuthbert did in this instance.

12: The Mentor

Raye faced the double-whammy of being Black and female, which was being written and talked about at that time. Although the women's movement was in full swing, many Black women did not see or feel that they were represented in it. More than anything, they knew they were fighting for their rights as women, but mostly as Black people. There are some interesting distinctions in this, and we consulted the following pieces for background: Frances Beal's 1969 pamphlet, "Double Jeopardy: To Be Black and Female," which was revised and published in Toni Cade Bambara's *The Black Woman: An Anthology* (New York: New American Library, 1970); Charlayne Hunter, "Many Blacks Wary of Women's Liberation Movement in U.S.," *New York Times*, November 17, 1970, p 47.

Paige Bowers spoke with Trenita Russell at length on January 7, 2019. The wide-ranging conversation provided a rich look at how Raye really did practice what she preached about mentoring people and empowering them to excel. Russell's story, in and of itself, is worthy of its own book, and her thoughts and memories of the NAVSEA offices were so valuable and entertaining.

13: David

Ultimately, this is a story about the power of a mother's love for her child. Up to this point we've seen how Flossie empowered Raye and how both of them worked hard to raise David to be a fine young man. For Paige, it was interesting to see how this maternal guidance shaped David as he went out on his own for the first time as a student at Morehouse College. Paige has come to know David as a highly educated, responsible, classy individual. Here, in this chapter, we see him as a young man who's a little bit lost but finding his way. David spoke with Paige about his undergraduate experience in Atlanta on July 19, 2019, and October 1, 2019. In the Raye J. Montague private archive, there is a photograph of Oprah Winfrey at the graduation ceremony, as well as several proud mom, Aunt Gladys, and Flossie pictures with David. There are also several letters from Flossie and Raye to David, which provide an endearing look at how much they missed him and wanted to fill him with news of home.

A Morehouse regional alumni meeting program shows that the school did honor Raye for her contributions to the school, among them setting up its Navy ROTC chapter.

14: On the Shoulders of Giants

Raye kept several of her speeches, including the undated "Communications and Image" that is referenced in this chapter. She was an avid speaker and member of the local Toastmasters chapter. Her involvement with Toastmasters was a carryover from her debate team days in college. She was always looking for ways to keep her public speaking and presentation skills sharp.

David Montague spoke with Paige Bowers on October 1, 2019, about the Links Inc. and how they were openly speaking about the importance of mental health at a time when it was taboo.

David spoke with Paige about his mother's after-work stress eating on July 19, 2019.

15: Retirement

David spoke with Paige about his mother's last year of work on multiple occasions, among them July 19, 2019, September 27, 2019, and October 1, 2019.

The article about the user benefits of CAD/CAM is in the Raye J. Montague private archive. Raye's personnel papers indicate that her comments were being hashed over by her boss Bill Tarbell and some of his other subordinates. Industry people wrote him about the unfortunate reality of the cost of CAD/CAM being out in the press. Ultimately this was becoming a project within the military that was useful, but challenging to find support for, because of its cost.

Bob Morgan counseled Raye on what to do after a department reorganization in a letter dated January 5, 1989.

Using Morgan's advice, Raye sent multiple memos to her bosses about what she was capable of doing after the department reorganization. The dates of these memos, each with different ideas of how her energy and talent could best be spent, are fairly close together. It is unknown what she was told about her ideas for herself during the restructuring, but one imagines she must have been told no to the first two fairly quickly and then been forced to adjust her course of action. It had to be anxiety inducing for Raye to be forced to find a place for herself, after having a niche across decades of consistent hard work. Those memos are in the Raye J. Montague private archive.

The flag that flew over the Capitol, replica canon, and signed caricature of Raye Montague are held in the Center for Arkansas History and Culture. The guest list for the retirement luncheon is held in the Raye J. Montague private archive.

16: Return to Little Rock

David Montague spoke with Paige Bowers about the last years of his mother's
    life on July 19, 2019.

Sandra Howell spoke with Paige Bowers on January 14, 2019.

Bonnie Dedrick spoke with Paige Bowers on January 15, 2019.

Lula Brooks spoke with Paige Bowers on April 11, 2019.

Boston "Baked Beans" Torrence spoke with Paige Bowers on January 7, 2019.

For details on the sailbot *Raye*, we consulted the article "Sailbot and MECH
    Celebrate Ada's Return and Future Endeavors," which can be found at the
    University of British Columbia's website: https://mech.ubc.ca/2018/03/10
    /sailbot-and-mech-celebrate-adas-return-and-future-endeavors/.

David and Paige visited the Arkansas Department of Corrections' Pine Bluff
    unit on July 18, 2019, and spoke with Vonnie Moore-Shabazz, who walked
    us around the prison grounds and showed us his cell and office. More than
    speaking about his affection for Raye Montague, he told us a story about his
    life. Born in Arkansas, he grew up in Los Angeles and was initiated into the
    Crips at age thirteen. When his mother moved back to Arkansas, she sent
    for Vonnie and his brother. Vonnie played football in high school, and then
    for Coach Tom Osborne at the University of Nebraska. After college, he
    joined the navy, got addicted to crack cocaine, and then his troubles with the
    law started. He received a life sentence in 1996 for robbery under the state's
    habitual offender law.

Donna Terrell spoke with Paige Bowers on January 11, 2019.

Rhonda Owen, "Raye Jean Jordan Montague: Confidence and ambition took
    Raye Montague from segregated schools to engineering breakthroughs as a
    civilian in the US Navy. She also taught herself to drive," *Arkansas Democrat-
    Gazette*, December 16, 2012, p 39.

"Meet the Woman Who Broke Barriers at the U.S. Navy," *Good Morning
    America*, February 20, 2017.

"Leading Lady: Raye Montague," *Harry*, October 12, 2017.

For information on the Silas Hunt awards, please see: https://diversity.uark.edu
    /get-involved/silas-hunt-awards.php.

Debra Moore-Lewis spoke with Paige Bowers on May 7, 2019.